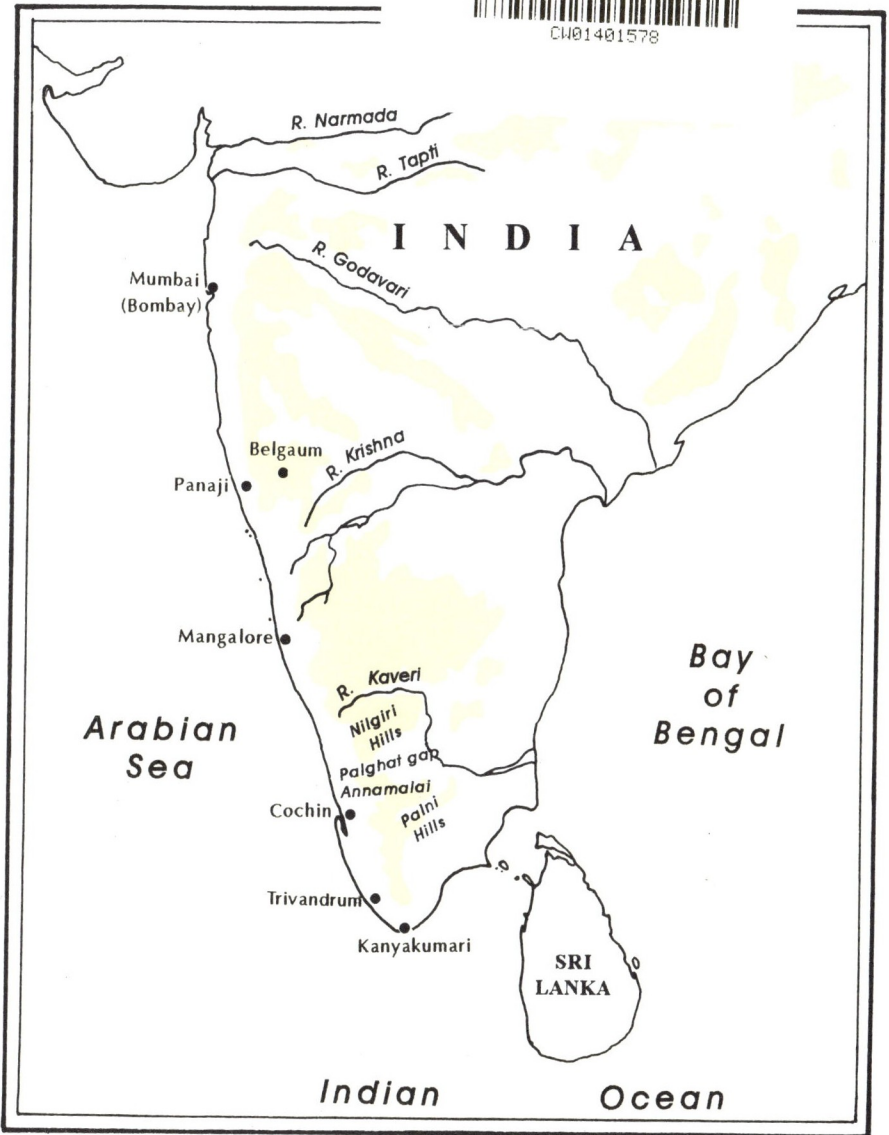

A
FIELD GUIDE
TO THE BIRDS OF
SOUTHWESTERN INDIA

A
FIELD GUIDE
TO THE BIRDS OF
SOUTHWESTERN INDIA

R.J. RANJIT DANIELS

DELHI
OXFORD UNIVERSITY PRESS
BOMBAY   CALCUTTA   MADRAS
1997

Oxford University Press, Walton Street, Oxford OX2 6DP

Oxford   New York
Athens   Auckland   Bangkok   Bombay
Calcutta   Cape Town   Dar es Salaam   Delhi
Florence   Hong Kong   Istanbul   Karachi
Kuala Lumpur   Madras   Madrid   Melbourne
Mexico City   Nairobi   Paris   Singapore
Taipei   Tokyo   Toronto

and associates in
Berlin   Ibadan

ISBN 0 19 563726 7

Typeset by SJI Services, B-17 Lajpat Nagar Part 2, New Delhi 110 024
Printed by Sahara India Mass Communication, Noida
Published by Manzar Khan, Oxford University Press, YMCA Library Building,
Jai Singh Road, New Delhi 110001

To
My father David DW Daniels
who helped me to begin
and to
My daughter Rosella
who will complete
what I have begun

CONTENTS

CONTENTS

ACKNOWLEDGEMENTS

This book is the culmination of twenty-five years of my self-tutoring in identifying and illustrating birds. My parents, brothers and sisters were a constant
source of encouragement throughout this mission. The past three years were
the most hectic when the text and illustrations were being prepared. During
this time my wife Vinetha and daughter Rosella co-operated beyond measure.
My early days in the forests of the Southern Western Ghats were made ad-
venturous and delightful, thanks to my old friend R.A. Edwin. Several others
have helped in many different ways. The remarks made by the late Dr. Sálim
Ali on some of my early illustrations made me feel confident while pursuing
my interest in birds. My career as a professional ornithologist was flagged
off in 1983 by the timely help of my good friend Yateendra Joshi. He introduced me to N.V. Joshi who in turn put me in touch with Professor Madhav
Gadgil at the Indian Institute of Science, Bangalore. Madhav and Sulochana
Gadgil were not only encouraging but also instrumental in publicizing my
bird illustrations in various ways. Shri Zafar Futehally has also always been
encouraging. The Bombay Natural History Society kindly permitted me to go
through the bird skins in their museum in 1983. Dr. V.S. Vijayan helped me
in finding papers on birds of Southwestern India published in century-old
journals such as *Stray Feathers*. While in the field, the forest departments of
Kerala, Tamil Nadu, Karnataka, Goa, and Maharashtra were of immense help.
R.B. Harikantra and M.J. Masti were faithful field assistants while working
in Uttara Kannada. Several people helped me while in Uttara Kannada. The
Ballarpur Salt Works kindly allowed me to spend days observing migratory
waterfowl and waders in their private salt pans. Stig Toft Madsen accompanied
me during some of my field trips and gave me many useful tips in the field
identification of waders. He also provided me with valuable literature on the
flight identification of raptors. The birdwatchers of Kerala provided me with
opportunities to explore the Silent Valley National Park. The National Institute
of Oceanography, Goa took me over to the Lakshadweep on one of their
annual cruises where I expanded my knowledge of seabirds. The Rev. K.N.
Matthews and those at the Sacred Heart College, Kodaikanal, kindly let me
study the birds in the Anglade Institute's Museum. Tim Inskipp provided me
with the revised list of names of Oriental birds. Ram Guha introduced my
work to the Oxford University Press. Shri Theodore Bhaskaran gave me some
of his notes on the early history of ornithology in India. The text was prepared
while in the Centre for Ecological Sciences, Indian Institute of Science. The
illustrations were made while resident at the Madras Crocodile Bank. Harish
Goankar, Christine Schonewald-Cox, and Prabhakar presented me with the
necessary art material. Numerous others have been of encouragement at
various stages. While I am not able to list each one here, I am grateful to all.

INTRODUCTION

With the growing popularity of birdwatching both as a hobby and as a science in India, it is appropriate to have field guides to the birds of smaller regions rather than of the country as a whole. A hundred years of ornithology in an area as small as Southwestern India has generated a considerable amount of information which goes far beyond mere checklists. The origin and patterns of distribution of the avifauna of this region are emerging. Community interactions, habitat choice and factors influencing the diversity of bird species in the southwestern hills and the adjacent plains have at least partly been understood. Efforts are being made to preserve the rarer and unique species of birds in Southwestern India. Recently, there has also been more interest in teaching the local human communities to conserve their avifauna. The present work is therefore a summary of all published information regarding the ornithology of Southwestern India during the past 100 years and a guide to those interested in studying the birds of this region in the field.

Southwestern India

Southwestern India includes the Western Ghats, the West Coast and a narrow strip of the Deccan Plateau. Biogeographically, the Western Ghats are the most important landscape features in Southwestern India due to their topography, rainfall, age and the resultant diversity of habitats. The Western Ghats form a continuous chain of tropical hills which runs between Kanyakumari in the south (8° N), to the River Tapti in the north (21° N), a distance of 1600 km. It is, however, broken at Palghat (11° N) by a pass 25 km wide. This gap is very significant in the biogeography of Southwestern India as it isolates the hills in the south from those in the north and links the flatter west coast with the eastern plateau, at least locally.

The hilly area defined as Southwestern India begins from the Arabian Sea in the west, goes over the rugged Western Ghats and merges with the Deccan Plateau in the east. The exact limits of this region cannot, however, be easily defined as the hills spur off eastwards in the Surat Dangs (Gujarat), Nilgiris (Tamil Nadu), Palnis (Tamil Nadu) and Biligirirangan Betta (Karnataka). Further, in parts of northern Karnataka (Karwar), the hills rise rather abruptly from the sea. Therefore, including the narrow coast, the ghats and the eastern spur-offs into the plateau, Southwestern India may be considered as spreading over approximately 160,000 km^2; an area that more or less coincides with the Malabar biogeographic province of India.

Biogeographers have divided Southwestern India into 4 smaller sections based on geology, climate and patterns of distribution of plants and animals. They are (1) the Northern Sahyadris; (2) the Southern Sahyadris; (3) the Nilgiris; and (4) the Southern Western Ghats. These divisions are appropriate in the context of birds as well. The hills and plains north of Goa, including

Maharashtra and Gujarat, are considered as the Northern Sahyadris. The hills in this section are younger, of a different geologic formation, and comparatively devoid of natural vegetation as against the rest of Southwestern India.

From Goa up to the Nilgiris, the hills are lower with a very narrow coast over much of their range. This section is called the Southern Sahyadris and is transitional between the rather unique species-rich Nilgiris and Southern Western Ghats, and the younger Northern Sahyadris. For instance, plants such as *Dipterocarpus indicus* and *Pinanga dicksoni* have their northern limits of distribution in this section as, do animals including many species of amphibians, the Travancore tortoise, the Wynaad laughing thrush and the liontailed macaque. Further, the yellowbacked sunbird found in the Northern Sahyadris is scarce south of this section.

The Nilgiris (including the Wynaad Plateau in Kerala) form the third section due to their complex topography and isolation from the other higher ranges. The mountains in this section are some of the oldest geological formations in Southwestern India.

South of the Palghat gap, the hills are unique and were earlier considered as part of an Indo-Ceylonese biogeographic province. The Palghat gap has isolated species of animals and plants from the Sahyadris and Nilgiris. Localized races of birds such as the whitebreasted laughing thrush, rufous babbler and blackbird are found in this section. The only extant population of the redfaced malkoha that we know of in India is restricted to this part of India, the other being in Sri Lanka. Moreover, in its climate this section is close to being equatorial with the most widespread rainfall. This section is therefore considered separately as the Southern Western Ghats.

The coast and the eastern plateau, being continuous, do not show any apparent variation in fauna or flora over the entire length of Southwestern India.

Topographically, Southwestern India is very varied. The terrain includes flat coastal areas in the west, the hilly Western Ghats like a backbone immediately to the east, and the flatter, though elevated eastern Deccan Plateau. The hills are gentler, rarely rising over 1300 m in the northern half, while they are steeper and higher in Kerala and Tamil Nadu. Peaks above 2000 m are, however, restricted to the Nilgiris, Palnis and Anaimalais. Exceptionally high peaks rise above 2500 m, the highest being Anaimudi Peak (2695 m) in the Anaimalais.

Climatically, Southwestern India is unique. Though the temperature is generally warm along the west coast, at about 100 km eastwards, it can drop to freezing point during winter. This is particularly common in the higher hills, especially the Nilgiris. It is hottest during April–May when the temperature goes beyond 30°C over most parts of Southwestern India. Generally, the coldest months are December–February. However, the coldest month locally south of 12° N is July which is also the wettest month.

Southwestern India gets much of its rain during June–October when the south-west monsoon is active. However, the southern parts get pre-monsoon showers beginning in April–May. The southern parts also receive rain during November–December when the north-east monsoon is active along the east coast of the peninsula. Thus, in and around the Nilgiris and further south, there are practically just 2–3 dry months, while in the north the dry season ranges from 4–5 months in the Southern Sahyadris to 5–8 months in the Northern Sahyadris. The length of the dry season has played a significant role in the present patterns of distribution of plants and animals in Southwestern India.

On an average the region receives 2500 mm of rainfall. This varies considerably, being higher along the western slopes of the Western Ghats and dwindling towards the eastern plateau. Heavy rainfall of over 5000 mm is common north of the Palghat gap, occasionally going beyond 7000 mm as in Agumbe (Karnataka) and 9000 mm in coastal Maharashtra. The lowest rainfall is around 1000 mm in the rain-shadow regions of the Palnis, Biligirirangan Betta, Bandipur and Mudumalai ranges and all along the eastern plateau. High rainfall and the undulating topography of the Western Ghats have rendered this narrow region humid with an abundance of perennial streams and rivers flowing both eastward and westward.

Vegetation

Plant life in Southwestern India is rich including at least 5000 species of flowering plants. The patterns of spatial distribution of these plants in relation to the climate and topography have given rise to broad, though distinct, vegetation types such as tropical evergreen rainforests, montane evergreen forests, tropical moist and dry deciduous forests, scrub, montane grasslands and tidal or mangrove forests. Semi-evergreen and secondary deciduous forests are found in the more seasonal Sahyadris and where human interference has been prolonged. Where human interference has been very severe, what remains is a kind of moist scrub-thicket with thorns and grass. Large areas on the Western Ghats have undergone total transformation of forests into plantations of coffee, tea, rubber, arecanut, coconut, teak and many other exotic tree crops such as pine, wattle and eucalyptus. Considerable extents of lush valleys in the Western Ghats have been usurped by massive hydroelectric projects.

Evergreen rainforests are restricted to the Western Ghats though they are no longer virgin or pristine. All evidence from records of forest history and direct study of vegetation in the field point to the fact that most, if not all, of the forests in the Western Ghats have been disturbed by man at one time or the other during the past 3000 years or so. Hence the evergreen rainforests found on the steeper slopes and in some deep valleys such as the Silent Valley (Kerala), Kodachadri (Karnataka) and parts of Agastyamalai (Tamil Nadu) are probably the least disturbed forests we have today; the rainforests in the

Silent Valley National Park apparently being some of the oldest forests in Southwestern India.

The evergreen rainforests of the Western Ghats can further be divided into low elevation, medium elevation, high elevation and montane rainforests. While the low, medium and high elevation forests do not differ considerably in floristic composition, the montane forests do. The montane rainforests are popularly called 'sholas' in India and 'elfin forests' elsewhere in the tropics. These forests are typically stunted with fewer species of trees and a predominance of mountain-adapted species such as *Rhododendron* occurring at elevations of above 1500 m in Southwestern India, bordered by grasslands, and occasionally, waterlogged swamps.

The evergreen rainforests of lower elevations are characterized by tall trees growing up to 50 m and many species of palms and lianas. Trees in the genera *Dipterocarpus, Garcinia, Myristica, Poeciloneuron, Cullenia, Canarium, Calophyllum, Diospyros, Persea, Mesua* and *Holigarna* tend to dominate in the vegetation. While species of *Garcinia* and *Myristica* are generally understorey trees, the others reach the canopy and beyond. These forests are also characterized by the presence of cane under the canopy and bamboo and reeds along the edges and streams. Torrential streams flow through these forests all through the year. Further, in many low-lying areas, swamps are created with a unique formation of stilt-rooted trees such as *Myristica magnifica* and the palm *Pinanga dicksoni*.

Deciduous forests are simpler in structure with fewer species of trees. Some of the best deciduous forests are found in and around the Mudumalai, Bandipur, Badra and Dandeli Wildlife Sanctuaries in Tamil Nadu and Karnataka. Moist deciduous forests occur in flatter areas receiving between 1500 and 2000 mm rainfall and are often richer in plants than the dry deciduous forests, which only get 1500 mm rain or less. However, these forests have many species of plants in common, being dominated by species of *Terminalia, Lagerstroemia, Dalbergia* amd *Albizzia*. Thorn, bamboo and tall grasses form the undergrowth.

Disturbed evergreen forests are often invaded by trees such as *Terminalia* and *Lagerstroemia* from the deciduous forests. These forests are considered as semi-evergreen or semi-deciduous. Such vegetation is more common in the Sahyadris where there is a longer dry season. In extreme cases secondary forests which are entirely deciduous have replaced the original evergreen forests in localities experiencing a long dry season. A good example of this may be seen in the Peechi Wildlife Sanctuary in Kerala.

Scrub jungles are found in the drier eastern foothills before the Western Ghats completely merge with the Deccan Plateau. The rainfall in these areas is often less than 1000 mm. Extensive scrub jungles are found in Moyar (Nilgiris), the eastern foothills of the Palnis, the Biligirirangan hills and Aramboli–Mahendragiri–Kalakad hills in southern Tamil Nadu. The vegetation is dominated by thorny acacias, cacti and grasses.

Along the mouths of all the larger west-flowing rivers are found tidal and mangrove forests. These are dominated by species of *Rhizophora, Avicennia* and *Sonneratia*. However, much of this vegetation is lost and what is left may be seen today in patches. The best and most extensive patches are available in Goa along the river Mandovi and in parts of coastal Karnataka. Extensive cultivation of coconut has considerably transformed the original coastal and estuarine landscape. Littoral vegetation characteristic of the beaches is also largely destroyed. Most of the beaches are inhabited by humans and maintain very little of the original vegetation and habitats.

One of the earliest and extant forms of tree cultivation in the Western Ghats was that of arecanut (*Areca catechu*) in the well-watered valleys. This age-old practice has created a rather stable, though secondary, ecosystem over the past 250 years. Monocultures of teak were raised much later during the 19th century. However, the more recent introduction of exotic tree species, such as wattle, pine, cinchona, tea, coffee, eucalyptus, and rubber, have considerably transformed the vegetation of the Western Ghats. The hills such as the Palnis and Nilgiris where the British colonists stayed have been affected the most with many introduced species becoming quite naturalized. Pine and eucalyptus were introduced in these hills as early as the 1830s and are still major elements of the landscape of the Western Ghats.

Shifting cultivation was practised on the Western Ghats until the end of the 19th century. Open grass and scrub-covered areas along the hills are telltale signs of this human occupation as are archaeological remains. Other areas have been converted for hill rice cultivation, and a constant demand on fodder, foliage and fuel for sustaining humans and their livestock has left much of the hills exposed and barren or covered with low thorny thickets and scrub.

Avifauna

A total of 508 species of birds in 599 forms have been hitherto reported from Southwestern India. These include 144 (28 per cent) species of water birds— ducks, herons and allied species, rails, seabirds and shorebirds. The avifauna of Southwestern India, despite being apparently rich, has a few rather unusual features. First, there are only 234 (64 per cent) species of the avifauna resident, 15 (3 per cent) being endemic. Second, the avifauna is dominated by non-passerine birds. There are only 195 species (38 per cent) of passerine birds in Southwestern India. This is certainly peculiar as in the entire world of the 8800 species of birds, 5200 (59 per cent) are passerines, and in the Indian subcontinent as a whole out of 1237 species 665 (54 pert cent) are passerines. While the Indian subcontinent conforms to the general global pattern, Southwestern India does not. Third, 1021 species (82 per cent) of the avifauna in the Indian subcontinent are land birds, whereas in Southwestern India there are only 363 (72 per cent) species of land birds.

The hills of Southwestern India, despite being tropical, are also poorer in bird species when compared with other parts of the tropics. Let us, for instance, consider the diversity of birds in tropical America. It is amazing that even a few square kilometres of forests in that continent have 200–400 species of resident birds. Nowhere in Southwestern India can we find such diversity. Even the extensive Silent Valley National Park with about 90 km^2 of well-preserved and mature rainforests does not harbour more than 200 species of birds including migrants.

The avifauna of Southwestern India is not only poor in terms of the number of species of birds but also lacks several guilds of birds unique to the tropical rainforests. For example, we do not find in the avifauna of Southwestern India guilds of such species as the antbirds of tropical America or the birds of paradise of New Guinea. Further, many families of birds characteristic of the tropical rainforests are represented by a single or just a few species. Examples may be quoted of the 12 species of pittas which are found in Southeast Asia as against the single species in Southwestern India. Similarly, we have just one species of trogon and one species of frogmouth while there are not less than 5 species in each of these families in Southeast Asia. There are only 5 species of sunbirds in Southwestern India whereas 76 and 24 species represent this family in Africa and Southeast Asia respectively.

Comparing Southwestern India with tropical America is probably not strictly valid as it has now been widely accepted that the rainforests of tropical America have been the largest centre of diversification of land birds and as a result they harbour an 'excess' of bird species. For instance, in the guild of pollinating birds alone there are 320 species of hummingbirds in tropical America as against the 116 species of sunbirds in the entire tropical Old World. These extra species have evolved over the past 1.5 million years when the earth was going through phases of cold and warmth due to the pleistocene glaciations. Theories suggest that tropical forests shrank into small patches during the cold periods and as a result birds got isolated for periods long enough to evolve into new species. When the warm climate returned, these isolated patches of forests merged and formed a large unit with a very diverse avifauna. The theory of pleistocene glaciation and diversification of birds in tropical America has been extended to Africa and other tropical countries. However, there is very little evidence that any of these areas outside tropical America had the pleistocene 'refugia'. Even in Africa, where the possibility of such refugia having existed has been shown, they are comparatively too few. Hence the avifauna in the Old World tropics has not diversified to the same extent as in tropical America.

The Eastern Himalayas are comparable to Southwestern India in extent of land and size of avifauna. There are 536 species of birds in the Eastern Himalayas of which 523 (98 per cent) are land birds. However 494 species (92 per cent) are resident which is much higher than the proportion resident in Southwestern India (64 per cent). The Eastern Himalayas have many more species of forest understorey birds when compared with the rainforests of

Southwestern India. In general, all families except the water birds in South-western India, are represented by more species in the Eastern Himalayas.

The high avian species diversity in the tropics is largely contributed by rainforests. Rainforests are also structurally the most complex of forests. They have more diversity of plant species, a greater vertical stratification, and denser canopy. This in turn creates more microhabitats and niches. It has been observed over the years and generalized that the more complex the forests, greater is the diversity of birds which they support. This general pattern observed in many of the tropical rainforests worldwide does not seem to fit the avifauna of Southwestern India. First, there are not many species of birds exclusive to the rainforests of this area. Second, the most structurally complex rainforests rank lower than the less complex deciduous and secondary forests in numbers of bird species, when equal areas are compared. How can this anomaly be explained?

The theory of island biogeography proposes that any region has a diversity of bird species determined by its size and distance from a larger source pool. These factors determine the type and number of species that invade and colonize a new area. Thus larger areas tend to have more species, as do those closer to the source pool. Proximity and larger area also tend to favour the colonization by specialized habitat users and birds with poorer dispersal capabilities. This phenomenon has been clearly demonstrated in many oceanic islands such as those around New Guinea. Isolated regions on continents can also behave like islands. The diversity and composition of avifauna in such isolated regions is determined by the distance across which colonizers can disperse, and also the available area which would be sufficient to support habitats for the immigrants.

Before we apply this phenomenon in trying to explain the pattern of avifaunal diversity observed in Southwestern India, we must look into the origin, prehistory and biogeography of birds in this region. Southwestern India was part of the Gondwanaland 150 million years ago. There is, however, no evidence that any species of bird existed in this region at that time. Bird life in Southwestern India apparently came into existence only after India merged with the Asian mainland about 45 million years ago.[1] All the families of birds represented in this narrow region today testify that our avifauna was largely derived from Africa and Southeast Asia. Thus we have no endemic families of birds.

It is not clear how these birds reached Southwestern India. Water birds in general are good dispersers, and as they spread over the globe they probably trickled into the peninsula too. But the land birds have apparently adopted different strategies. The alternating arid-cold and humid-warm climates that the entire tropics experienced during the pleistocene period are believed to

---

[1] There is some recent evidence from the Siwaliks of Pakistan suggesting that this collision took place around 20 million years ago.

have favoured periodic invasions of both forest and montane birds into South-western India. The hilltops of the Satpuras and Eastern Ghats have certainly played a role in the movement of birds in the past. Geologic evidence suggests that a maximum number of species invaded the peninsula from Africa and Southeast Asia 250,000 years ago. During this period, birds that successfully colonized newer areas started diversifying locally while the unsuccessful species were getting eliminated. This process probably continued until about 10,000 years ago, when the last arid-cold period withdrew from the tropics. The early colonizers evolved into full species. The latecomers have just dif-ferentiated into endemic races or subspecies. As a result, as is apparent, more latecomers than early arrivals have survived in Southwestern India. Thus, while there are only 15 endemic species, there are over 50 races exclusive to this region.

The patterns of distribution of some of the resident birds of Southwestern India are worth discussing further. To begin with, of the endemic species, the Malabar parakeet, whitebellied tree pie, greyheaded bulbul, rufous babbler, whitebellied blue flycatcher, Malabar whistling thrush and crimsonbacked sun-bird are rainforest birds of lower elevations and are also widespread in South-western India. In contrast, the Nilgiri wood pigeon, black and rufous flycatcher, Nilgiri flycatcher, rufousbreasted laughing thrush, greybreasted laughing thrush, whitebellied shortwing and Nilgiri pipit are not strictly rain-forest birds. However, these birds are montane birds, and, except for the wood pigeon, all are restricted to the hills in the south. Another exception to the general pattern is the Malabar lark. This species is neither a forest bird nor a montane species. It is a bird of the open plains and higher grasslands in Southwestern India. Moreover, it is the most widespread of the endemic birds both in geographical range and altitude. It is also interesting to find that the races of birds isolated from the northeastern rainforests in Southwestern India are largely of low-elevation species. Some examples are the tiger bittern, bazas, rufousbellied hawk-eagle, maroonbacked imperial pigeon and Wynaad laughing thrush. This suggests that montane and grassland birds were isolated on the Western Ghats at a much earlier period than the birds of the lower-elevation rainforests and opener habitats. Recent geological evidence also points to the fact that the cold-arid periods lasted longer during the pleistocene than the warm-humid periods. This may also explain why the birds of the low-elevation rainforests in Southwestern India have not differentiated much.

Due to the gradual warming of the tropics during the past 10,000 years and the increasing population of human beings in peninsular India, the extent of rainforests in Southwestern India has shrunk considerably. During the past 3500 years, particularly after the invention of iron tools, the rainforests have disappeared more rapidly, giving way to grasslands and scrub. Subsequently, a greater number of birds adapted to the scrub, and deciduous and secondary forests spread far and wide, even invading the more humid zones along the Western Ghats, limiting the immigration and success of the rainforest birds.

The patchy distribution of undifferentiated species of rainforest birds in the Western Ghats, Sri Lanka and the Eastern Himalayas suggests that low levels of immigration and higher rates of local extinction in the recent past have clearly determined the present avifauna in this part of the tropics. Examples may be quoted of the Ceylon frogmouth and great-eared nightjar in the Western Ghats with isolated populations of the species in Sri Lanka and the Eastern Himalayas respectively. Others such as an extant population of Nilgiri wood pigeons in the Nandi hills, about 300 km to the east of the nearest hills in the Western Ghats, and the records of the black bulbul and whitebellied tree pie in 1922 from Maharashtra, about 100 km north of their present ranges, are evidence that the rainforests are continuously shrinking, locally exterminating or isolating populations of birds.

Poor dispersers such as babblers and laughing thrushes, which are important forest understorey passerines in the Indo-Malayan rainforests, have not successfully colonized the hills of Southwestern India due to the great barrier of 1500 km-wide dry forests and scrub in the peninsula. On the contrary, they have invaded the contiguous Eastern Himalayan rainforests, diversified locally and contributed significantly to the diversity of avifauna in the northeastern hills. A detailed comparison with the avifauna of the Eastern Himalayas further suggests that the Western Ghats are particularly impoverished in species of rainforest babblers and laughing thrushes. Excluding those species found above 2500 m in the Eastern Himalayas, we still find 56 species of rainforest babblers and laughing thrushes that do not occur in Southwestern India. The absence of these species (they make up more than 10 per cent of the Eastern Himalayan avifauna) has largely reduced the diversity of birds in Southwestern India in general and the rainforests in particular.

It is now apparent how the rainforests of Southwestern India derived their avifauna. With no further scope for receiving more species from a distant source pool such as the eastern Indo-Malayan rainforests, the avifauna of the rainforests of Southwestern India has been saturated at a lower level of diversity. On the contrary, the deciduous forests of Southwestern India, being contiguous with a more extensive vegetation spread over the entire country, are directly under the influence of a comparatively larger, though commoner and more widespread, pool of bird species. Hence they tend to have a greater number of species of birds than the more complex rainforests in Southwestern India.

History of ornithology
The history of ornithology in Southwestern India falls into two distinct periods: the pre-independence British period and the post-independence period. The British period lasted for 75 years between 1860 and 1935. After this period, the late Dr Sálim Ali entered the scene, and Indian field ornithology came into its own.

The British period can be clearly divided into three distinct eras. The first era (1860–72) was that of T. C. Jerdon. This era certainly marks the beginning of ornithology in the Western Ghats. Excellent British ornithologists such as Edward Blyth and Brian Hodgson were in India during this period. Jerdon, Blyth and Hodgson are considered the founders of Indian ornithology. Their contribution to the knowledge of our birds is memorable. During this period, some coffee planters in the Western Ghats were also actively studying birds. Important among them were T. F. Bourdillon, his brother H. T. Fulton, and H. S. Ferguson. Bourdillon was also the Conservator of Forests in Travancore (Kerala). The ornithology of Southwestern India owes a lot to the simple beginnings made by these pioneers.

The second era was that of Allan Octavian Hume between 1872 and 1922. Hume, the 'Pope of Indian Ornithology', was also the founder of the Indian National Congress. It was during this time that ornithological studies in India began to take direction. Hume edited *Stray Feathers*, a journal dedicated to the study of birds. This journal however ceased publication in 1888.

The Bombay Natural History Society was founded during this second era in 1886. It was meant to bring together naturalists with their notes and collections under one roof. A good collection of bird skins have since then been lodged in the society's museum. The Journal of the Bombay Natural History Society which was subsequently released in 1886 filled the void left by *Stray Feathers*. Ornithological notes and lists of birds from different parts of the country were regularly published in the journal.

The first lists of birds for any part of Southwestern India were published in *Stray Feathers* in 1876 by the Rev. S. B. Fairbank for Sawantwadi (Maharashtra), and by Hume for Travancore. Preliminary lists of birds from the Palni hills were published by Fairbank in 1877 and by H. A. Terry in 1887. G. W. Vidal's list of birds for coastal Maharashtra appeared in *Stray Feathers* during 1880. This included the unpublished notes of Jerdon and James Armstrong who had worked briefly in this part of the Western Ghats. J. Davidson listed the birds of western Khandesh in Maharashtra in 1885. However, his most significant contribution was the list of birds made between 1888 and 1896 in the Uttara Kannada (North Kanara) district of Karnataka. His lists were published with details of distribution and breeding as two parts in 1898 in the *Journal of the Bombay Natural History Society*. The work of Davidson is monumental, including more than 340 species of birds and rendering the small district of Uttara Kannada the most completely surveyed with regard to birds in Southwestern India . The second era also marks the beginning of the study of bird eggs in Southwestern India. John Stuart, a planter, collected and catalogued eggs between 1888 and 1890. These eggs were to become part of a much larger collection of Indian bird eggs which led to the publication of *The Nidification of Birds of the Indian Empire* by E. C. Stuart Baker.

Towards the end of the second era more ornithologists came into the scene. The first books on Indian birds were also being published during this time. Most important among these was the first series of the *Fauna of British India—Birds,* compiled by W. T. Blanford and E. W. Oates. Hugh Whistler, who made the first attempt at preparing a popular bird guide, *The Popular Handbook of Indian Birds,* was active in India during this era.

The third era was brief. It also ended the British domination of Indian ornithology in general. It lasted between 1922 and 1940. During this period the most sigificant work with Indian birds was that of E.C. Stuart Baker. His most valuable contribution was the revised *Fauna of British India* series on birds (1922–30). These books are still some of the best morphological accounts of birds available to any student of Indian ornithology. A few other foreign ornithologists did study the birds of the Western Ghats during this period, but their studies were localized. F. N. Betts contributed significantly to our knowledge of the birds of Coorg (Kodagu) and also of the Nilgiris. R. S. P. Bates worked briefly in Coorg and the Nilgiris. A. P. Kinloch published his list of birds for the Nelliampathy Hills in Kerala. In 1938 another bird taxonomist, Walter Koelz, visited the northern parts of Uttara Kannada to collect specimens for the American Museum of Natural History. Within a short period of three months he collected 230 species of birds and published the list in the *Journal of the Bombay Natural History Society* in 1942. The preserved skins are at the American Museum of Natural History under the 'Koelz Collections'.

All through the British period, other agencies were indirectly contributing to the development of ornithology in Southwestern India. A number of European planters in the hills were periodically contributing their bird collections and notes to the Bombay Natural History Society. Ever since the Jesuit training centre, the Sacred Heart College, came into existence at Kodaikanal, the natural history of the Palni hills was explored. The college was first established in 1895 and the museum preserves, among others, skins of birds collected from the Palni Hills during the past century. Another body that needs mention is the Nilgiri Game Association founded in the Nilgiris in 1879. Records of game hunted and the attempts that were made by this association to introduce exotic game birds in the Western Ghats are informative.

The post-British period started with the late Dr Sálim Ali's survey of the birds of Travancore and Cochin in 1935. His later surveys of the birds of Mysore in 1939–40 and those of Goa in 1975 with Robert Grubh have considerably enhanced our knowledge of the birds of Southwestern India. Sálim Ali's cousin and one time associate, Humayun Abdulali, is well known for his meticulous taxonomic study of Indian birds. His lists of birds for the large state of Maharashtra, and that of Borivali National Park (1981) in particular are the most complete lists available for the Northern Sahyadris. Further contributions were made during this period by the Rev. Norman A. Fuller at the

Sacred Heart College, Shembaganur, after his brief study of the birds of the Palni hills.

The most popular and outstanding piece of work in the history of ornithology in Southwestern India is that of the late Professor K. K. Neelakantan of Kerala. Over the past 50 years his devoted studies of the birds in his state, besides his studies elsewhere, published both in English as well as in Malayalam, have been a boon to any birdwatcher in Kerala. It is beyond doubt that his illustrated guide to the birds of Kerala *Keralathile Pakshikal*, first published in 1958 in Malayalam, is the only standard book on birds available for any part of the country in a regional language.

Zafar Futehally's contribution to the study of Indian birds is remarkable. His interest in birds and conservation founded and maintained the *Newsletter for Birdwatchers* for the past 30 years. This monthly magazine dedicated to Indian birds has been one of the best means of communicating ornithological notes to interested birdwatchers all over the country and abroad.

Since 1983, the Indian Institute of Science has been studying the birds of Southwestern India in general and those of the Uttara Kannada district in particular. The outcome of the research in Uttara Kannada district is probably the most complete study of the community ecology and species diversity of birds for any part of Southwestern India. Earlier attempts were made on these lines by V. S. Vijayan and Lalitha Vijayan at the Periyar Wildlife Sanctuary. Reza Khan has studied the birds of the Nilgiri sholas and other montane habitats. A. J. T. Johnsingh's work on the birds of the Mundanthurai Wildlife Sanctuary in the Southern Western Ghats also deserves mention.

Work on the birds of Southwestern India is being continued by amateurs as well as interested students. Two recent surveys organized by the birdwatchers in Kerala in collaboration with the local forest departments are worth mentioning. The first was carried out in December 1990 at the Silent Valley National Park. The second was conducted in February–March 1991, in Peechi Wildlife Sanctuary, Trichur. The latter was organized by the Nature Education and Science Trust of Trichur. More surveys and detailed studies will enhance our knowledge of the birds of this unique region of the world.

Watching birds

Watching birds in Southwestern India can be both delightful and frustrating. It depends largely on what one wishes to find and where, and also varies considerably with the season. To be 'up and out with the lark' is not necessary in this region, especially for those birdwatching in the Western Ghats. There are but few 'early birds' in these hills.

The most comfortable and rewarding season for watching birds in most parts of Southwestern India is between October and May. The rains cease to be adamant during October and it is also the time when all the migratory birds arrive. April and May are generally hot months, though they can be rainy in the south and in the Nilgiris. During these summer months, early morning and late evening walks would be very fruitful. Watching birds within dense forests is difficult though there are birds all through the day. A walk along the forest paths and roads along the edges of dense forests would be the most exciting. Any time between 6.30 and 8.00 am is appropriate for such outings. On a good day one might see at least 60 species of birds within two hours or so of walking in such open and secondary forest habitats.

February–March mornings are usually misty with a lot of dew falling. In these months, watching birds at dawn can be frustrating in most parts of Southwestern India unless one is familiar with all the bird calls and songs. Birds are particularly vocal in Southwestern India during this season. However, poor lighting and cold do not favour watching them. Raptors are late risers. They are mostly found perched on exposed treetops at dawn. Mere silhouettes of most raptors, even at close range, are next to useless in identifying them with certainty. Raptors are best identified in overhead flight and for this midday would be the best time.

Edges of open forests, village roads and dry streambeds are exciting and safe for a long evening walk. Dusk and early moonlit nights are ideal for watching nocturnal birds such as nightjars and owls. These birds are never really silent at this hour and are best identified by their characteristic calls.

October–December is an excellent time for watching water birds as would be April–May. The estuaries, salt pans and larger reservoirs are full of migratory water birds at the beginning of the season. Towards the end when the marshes dry up as summer approaches, large gatherings of storks, lapwings and wagtails may be found.

The hill towns of the Western Ghats such as Munnar (Kerala), Shembaganur and Kodaikanal (Palni hills), Ooty and Coonoor (Nilgiri hills), Valparai Estates (Anaimalais) and KR hills (Karnataka) are among the best places to watch montane birds such as laughing thrushes and flycatchers. Many species of forest birds are common in other towns at lower elevations too. The outskirts of Sirsi, Castle Rock and Dandeli in northern Karnataka, Ponda in Goa and Trichur in Kerala are a few excellent places for watching birds in Southwestern India. Human settlements on hills, arecanut plantations and all estates are easy terrain with a delightful avifauna.

For those interested in watching birds only in the wilderness away from human habitation, all the sanctuaries and national parks would be ideal. However, in Periyar, Mudumalai and Bandipur, there is a considerable movement of elephants all through the year so that birdwatching involves taking a risk. For the sake of safety, it is best to stay in the periphery of these wildlife preserves. Watching birds in the Silent Valley National Park is not as risky. The Dandeli Wildlife Sanctuary, the little known Peechi Wildlife Sanctuary in Trichur, the Molem Wildlife Sanctuary in Goa, the Borivali National Park in Mumbai, and the Mundanthurai-Kalakad Wildlife Sanctuaries are some of the best places for watching birds without the fear of encountering elephants.

There are a few very interesting spots for watching migratory water birds in Southwestern India. The best would be the large estuary and salt marshes of the River Aganashini south of Karwar in coastal Karnataka. Ducks and shorebirds gather in thousands along this estuary. Bird-life is not particularly diverse in mangrove vegetation. However, the little-seen blackcapped and whitecollared kingfishers are more certainly sighted in this habitat, the whitecollared kingfisher being extremely rare in Southwestern India.

Field Guide

This section describes the 508 species of birds identified up to now in Southwestern India. The descriptions include as many clues as possible that help identification of every species in the field. The clues include size, colour, general shape, unique behaviour (if any), call and habitat. However, whichever of these features is the most striking in the field, appears first in the description without any rigid sequence. Thus for each species, the first few lines of description are brief, pointing out the most diagnostic characteristics. Further descriptions include other details such as call, range and habitat. Whenever a species is difficult to identify in the field, clues to identifying in hand are provided. As far as possible, technical terms have been avoided in describing birds.

Classification and nomenclature

After the publication of *A Synopsis of the Birds of India and Pakistan* by S. Dillon Ripley in 1983, considerable revisions have been made in bird taxonomy and systematics. The most recent classification and nomenclature is that of C. G. Sibley and B. L. Monroe (1990) in *Distribution and Taxonomy of Birds of the World*. To be consistent with the international bird nomenclature, the common English names used by Sibley and Monroe have been followed in the present work. The more popular English names in use within India have nevertheless been given in brackets. Changes in Latin names have been incorporated without additional comments. The revised systematics of families and species given by Sibley and Monroe has, however, not been used as such in this book, since it is drastically different from all schemes in use until recently. However, wherever necessary, appropriate remarks have been made regarding changes at the level of orders and families. For the sake of simplicity, the family names and species code numbers used in all Indian bird literature and in the museum of the Bombay Natural History Society are maintained. A separate continuous serial numbering of species is, therefore, not included.

Identification

For easy identification in the field a brief description of the family precedes each set of related species. Except in the case of the large family MUS-

CICAPIDAE, no attempt has been made to describe the subfamilies. Family descriptions are general including broad colours, habits and habitats.

Species descriptions start with size, this being one of the best means of separating species. However, sizes are difficult to visualize in the field if there is nothing beside to compare with. Therefore, in addition to the simplest and precise measure of size, viz. the length of a bird from the tip of the beak to the tip of the tail in centimetres, relative measures such as 'small', 'large', 'tall', etc., are included. These measures are relative to the sizes of the smallest and largest birds within a family. Thus, a large warbler may actually be smaller than a drongo and the smallest hornbill is larger than a crow.

Colours are certainly the best indicators to the identity of a species at close range or through binoculars. However, colours are frequently very variable with age, sex and season. Many birds of prey and owls have more than one colour phase. This complicates identification in the field. Hence for those species with variable plumage and colours, details of all stages are given according to the frequency of their occurrence. Thus, if a bird is normally seen in winter colours in Southwestern India, the description first refers to this. Where sexes differ in colour, it is the male that is described first as males are more frequently noticed in the field. In birds, the unfeathered or bare parts also show characteristic colours. Though these may also vary with age, sex and season, in many species the colours of the eyes, beak, mouth, legs and feet are diagnostic.

Shapes of birds often help to identify species from a distance. Details such as 'long-necked', 'short-tailed', 'slender' and 'stocky', readily separate certain species in the field. Wherever diagnostic, it is best to go by such details, especially under conditions of poor lighting. Unique behaviour and habits are very useful clues in the field. Flicking wings or a wagging tail, fast flight, being noisy or silent, solitary or gregarious, are characteristics of some species.

For all species existing as more than one distinct race in Southwestern India, clues to separating the races both in the field and in hand are included. However, except in a few cases, races are not easily separated in the field. Hence it is best not to pursue this much in the field. Generally, the ranges are enough to separate the races which rarely overlap except in winter.

Calls are very important and often the only clue to identifying birds in the forests of the Western Ghats. This is particularly so while encountering nocturnal birds, species of cuckoos, warblers and birds of the forest interiors. While some calls can easily be described or rendered verbally, many are complex. Hence wherever possible and known, the most commonly heard calls and/or songs are described. Many calls are from the author's field notes. Those already described and rendered by others in the literature are indicated accordingly.

Range

Range is the geographic spread of any species in Southwestern India as we know it today. Species extend their ranges and sometimes become extinct locally. Ranges such as 'the entire Western Ghats' suggest that the species is known to occur from Kanyakumari to the river Tapti. A widespread species is one that occurs in many parts of Southwestern India over a fairly wide range. The species need not, however, be found everywhere within the range. Local species are known from very small areas such as the Nilgiri hills or south of Palghat. Species not known outside Southwestern India are endemic. Altitudinal ranges of most species are rather poorly known. However, as in the field it is often easier to separate closely resembling species by altitude than by other clues if definitely known, the lower and higher limits of the altitudinal range of a species are given.

Habitat

Any species of bird utilizes one or more habitats during its lifetime. This often varies with age, sex and season. Hence, for each species, all habitats in which it has been recorded are listed. The most frequently utilized habitat is, however, highlighted. As many closely related species are separated by habitats due to ecological reasons, a knowledge of the appropriate habitat of a species is very useful in the field.

Illustrations

Illustrations have primarily been made to readily help in identifying all the birds in the field. For the more difficult species, additional illustrations have been provided in black and white, as in the case of warblers where wing formulae have been shown and raptors for which flight silhouettes are often necessary. Detailed illustrations of juvenile, adult and seasonal plumages are not within the scope of this book and hence are not included except when absolutely necessary. Pointers have been used to help look for the right features in the field.

CLASSIFICATION OF THE BIRDS OF SOUTHWESTERN INDIA

Order	Family	Common Names
PODICIPEDIFORMES	PODICIPEDIDAE	Grebes
PROCELLARIIFORMES	PROCELLARIIDAE	Shearwaters
	HYDROBATIDAE	Storm petrels
PELECANIFORMES	PHAETHONTIDAE	Tropic birds
	PELECANIDAE	Pelicans
	SULIDAE	Boobies
	PHALACROCORACIDAE	Cormorants, Darters
	FREGATIDAE	Frigate birds
CICONIIFORMES	ARDEIDAE	Herons, Bitterns
	CICONIIDAE	Storks
	THRESKIORNITHIDAE	Ibises, Spoonbills
	PHOENICOPTERIDAE	Flamingos
ANSERIFORMES	ANATIDAE	Ducks, Teals
FALCONIFORMES	ACCIPITRIDAE	Hawks, Kites, Eagles Buzzards, Harriers Vultures, Ospreys
	FALCONIDAE	Falcons
GALLIFORMES	PHASIANIDAE	Partridges, Quails Peafowl, Junglefowl, Spurfowl
GRUIFORMES	TURNICIDAE	Bustard-quails
	GRUIDAE	Cranes
	RALLIDAE	Rails, Crakes, Coots
	OTIDIDAE	Floricans
CHARADRIIFORMES	JACANIDAE	Jacanas
	HAEMATOPODIDAE	Oystercatchers
	CHARADRIIDAE	Plovers, Lapwings
	SCOLOPACIDAE	Curlews, Sandpipers Snipes, Phalaropes
	ROSTRATULIDAE	Painted snipes
	RECURVIROSTRIDAE	Stilts, Avocets
	DROMADIDAE	Crab plovers
	BURHINIDAE	Stone curlews, Stone plovers
	GLAREOLIDAE	Coursers, Pratincoles
	STERCORARIIDAE	Skuas
	LARIDAE	Gulls, Terns
COLUMBIFORMES	PTEROCLIDIDAE	Sandgrouse
	COLUMBIDAE	Pigeons, Doves
PSITTACIFORMES	PSITTACIDAE	Parakeets, Lorikeets

CUCULIFORMES	CUCULIDAE	Cuckoos, Coucals Malkohas
STRIGIFORMES	STRIGIDAE	Owls
CAPRIMULGIFORMES	PODARGIDAE	Frogmouths
	CAPRIMULGIDAE	Nightjars
APODIFORMES	APODIDAE	Swifts
TROGONIFORMES	TROGONIDAE	Trogons
CORACIIFORMES	ALCEDINIDAE	Kingfishers
	MEROPIDAE	Bee-eaters
	CORACIIDAE	Rollers
	UPUPIDAE	Hoopoes
	BUCEROTIDAE	Hornbills
PICIFORMES	CAPITONIDAE	Barbets
	PICIDAE	Woodpeckers, Piculets, Wrynecks
PASSERIFORMES	PITTIDAE	Pittas
	ALAUDIDAE	Larks
	HIRUNDINIDAE	Swallows, Martins
	LANIIDAE	Shrikes
	ORIOLIDAE	Orioles
	DICRURIDAE	Drongos
	ARTAMIDAE	Swallow-shrikes
	STURNIDAE	Mynas, Starlings
	CORVIDAE	Crows, Tree pies
	BOMBYCILLIDAE	Hypocolius
	CAMPEPHAGIDAE	Minivets, Cuckoo-shrikes, Wood-shrikes, Flycatcher-shrikes
	IRENIDAE	Fairy bluebirds, Chloropses, Ioras
	PYCNONOTIDAE	Bulbuls
	MUSCICAPIDAE	Babblers, Flycatchers, Warblers, Thrushes
	PARIDAE	Tits
	SITTIDAE	Nuthatches
	MOTACILLIDAE	Pipits, Wagtails
	DICAEIDAE	Flowerpeckers
NECTARINIIDAE		Sunbirds
	ZOSTEROPIDAE	White-eyes
	PLOCEIDAE	Weaverbirds, Munias, Sparrows
	FRINGILLIDAE	Rosefinches
	EMBERIZIDAE	Buntings

PARTS OF A BIRD

1. Bill/Beak
2. Forehead
3. Crown
4. Eyebrow/Supercilium
5. Nape
6. Hindneck
7. Eye / Iris
8. Throat
9. Breast
10. Shoulder
11. Abdomen/Belly
12. Back
13. Flank
14. Rump
15. Vent/Under tail-coverts

16. Upper tail-coverts
17. Outer tail feathers
18. Wing length
19. Wing speculum
20. Secondary feathers/Secondaries
21. Wing-coverts
22. Trailing edge of wing
23 Primary feathers/Primaries
24. Wing-mirrors
25. Outer toe
26. Middle toe
27. Inner toe
28. Hind toe
29. Tarsus length

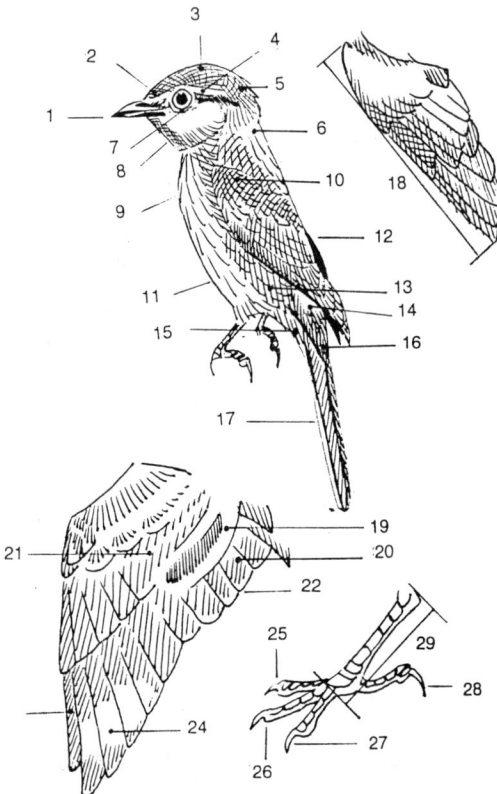

GLOSSARY

Endemic A population of birds restricted to just one part of the world. An identical population does not exist elsewhere. The population may belong to a full species or to a sub-species.

Guild Groups of organisms (birds) utilizing a given resource in a similar way. The resource may be food or habitat.

Microhabitat Space or substrate of smaller dimensions within the habitat itself. For example, if a forest is considered as a habitat, the forest floor may be a microhabitat. This, however, varies with the scale in which habitats are defined.

Niche The position an organism (bird) fills in an ecosystem. It is multidimensional as an organism can have a feeding niche, a habitat niche, a breeding niche, etc.

Nonpasserines Birds of all orders other than order PASSERIFORMES. Woodpeckers and all families lower in taxonomic hierarchy belong to this broad category.

Passerines The most evolved birds often referred to as songbirds. All families in the order PASSERIFORMES beginning with PITTIDAE.

Race Geographically isolated populations of a species which have differentiated genetically. Races however interbreed on contact. A named race is a subspecies.

Refugia Isolated patches of habitat within which a species survived during adverse climatic conditions.

Species Populations of organisms (birds) which breed amongst themselves but are reproductively isolated from other such groups.

FIELD GUIDE

Family PODICIPEDIDAE

Aquatic. Swimming birds with soft stumpy tails. Wings short. Beaks sharp and pointed. Downy young with black and white stripes. Sexes alike.

5.   Little Grebe *Tachybaptus ruficollis* (Dabchick)

Identification: 23 cm. A small, brown duck-like bird. Tail-less appearance diagnostic. Pairs or flocks. **Breeding**: Dark brown; crown darker. Head, face and throat chestnut. Smoky-white underparts with dark brown flanks. In flight, white patch on secondaries conspicuous. **Non-breeding**: Drab brown; paler below.
Call: Sharp squeaks and musical trills. Another .. *click* ...
Range: Resident with local movements. Entire Southwestern India.
Habitat: All kinds of wetlands, especially tanks and lakes with floating vegetation.

Family PROCELLARIIDAE

Seabirds of various sizes. Grey, black, brown, white or combinations of these. Wings narrow, long and pointed. Tail rounded. Nostrils tubular. Feet webbed. Sexes alike.

12.   Persian Shearwater *Puffinus persicus*

Identification: 30 cm. A medium-sized brown-black and white seabird. Darker underwing and under tail-coverts diagnostic. **Adult**: Sooty-black. Greyish neck; the grey extending to the sides of breast. White below. Under tail-coverts black and white or full black. Legs and feet pinkish-white.
Range: Stragglers along the West Coast. Wind blown specimens recorded on Mumbai and Kerala coasts.
Habitat: Marine.

Family HYDROBATIDAE

Smallest of seabirds. Black, grey and white. Wings long and pointed. Nostrils tubular. Legs slender. Feet webbed. Sexes alike.

14.   Wilson's Storm Petrel *Oceanites oceanicus*

Identification: 19 cm. A very small, dark, swift-like seabird with white rump. Feet projecting beyond tip of tail in flight. Flocks foraging over water in sea. **Adult**: Sooty-black with conspicuous white rump. In flight a pale upper wingbar diagnostic. Legs and feet black with yellow webs.

Range: Mumbai, Konkan coast and Kanyakumari.
Habitat: Marine. Occasionally strays closer to the coast.

Family PHAETHONTIDAE

Tropical tern-like seabirds with long ribbons in the wedge-shaped tails. Predominantly white. Red beak. Short webbed feet. In open sea, often freely swimming or resting on water.

18.   Redbilled Tropic Bird *Phaethon aethereus* (Short-tailed Tropic Bird)

Identification: 40 cm (+ tail 30 cm). A largish slender white tern-like bird of open seas. Long streamers from tail diagnostic. Black crescentic band from eyes around hindneck. Black wing-tips conspicuous in flight. Bright coral-red beak. **Immature**: Black spotting on upperparts. Lacks the tail streamers. Black nuchal crest and yellow beak diagnostic.
Range: Stragglers on the West Coast reported from Mumbai and southern Karnataka.
Habitat: Marine.

Family PELECANIDAE

Large, clumsy aquatic birds with short legs. Beaks massive with extensible pouch. Feet fully webbed. Tail short. Sexes alike.

21.   Spotbilled Pelican *Pelecanus philippensis* (Grey Pelican)

Identification: 152 cm. A very large short-legged water bird with massive beak. Grey colour diagnostic. **Adult**: Grey above. Crest on nape brown with white tips. Whitish underparts. Beak pinkish with blue spots; pouch purplish with darker spots. **Immature**: Pale brown above and white below.
Range: Stragglers along eastern Southwestern India.
Habitat: Large lakes, reservoirs and rivers.

Family SULIDAE

Large seabirds with stout, pointed beaks. Wings long and pointed. Tail and legs short. Feet webbed. Face, legs and feet brightly coloured. Sexes alike.

23. Masked Booby *Sula dactylatra*

Identification: 80 cm. A large white and black seabird with stout pointed beak. Plunges in water with short wedge-shaped tail fanned out. **Adult**: White with black wing feathers (primaries and greater wing-coverts). Beak bluish-grey, yellow or red. Naked skin on face blue-black. Legs and feet bluish, yellow or orange (never red). **Immature**: Head and neck dark brown; upperparts greyish brown.
Range: Wind blown specimens from coasts of Mumbai, Karnataka and northern Kerala.
Habitat: Marine.

Family PHALACROCORACIDAE

Sleek, colonial water birds with sharp beaks. Black. Neck and tail long. Legs short. Feet webbed. Swimming with only neck and head above the surface. Diving. Perch exposed with wings spread out. Sexes alike.

NOTE: Sibley and Monroe (1990) treat the Darter as belonging to family AN-HINGIDAE.

26. Great Cormorant *Phalacrocorax carbo* (Large Cormorant)

Identification: 80 cm. The largest cormorant. White and yellow patches on underparts contrasting with glossy black upperparts diagnostic. **Adult breeding**: Glossy black with white patches on flanks; head, neck and crest mixed with white. Bare face white and gular skin yellow. **Adult non-breeding**: Mostly black. Gular skin duller. **Immature**: Brown above progressively darkening with age; dull white below.
Range: Resident with local movements. Widespread in Southwestern India.
Habitat: Inland waters and estuaries.

27. Indian Cormorant *Phalacrocorax fuscicollis* (Shag)

Identification: 63 cm. A cormorant intermediate in size between large and little cormorants. Yellow gular skin and sloping forehead separate from little cormorant. **Adult breeding**: Glossy black with a scaly effect above. Jet-black below. Pure white tuft of feathers on sides of neck. **Adult non-breeding**: Black with traces of white on plumage. **Immature**: Brownish with scaly pattern on back. Whitish below.
Range: Resident with local movements. Widespread in Southwestern India.
Habitat: Lakes, rivers, reservoirs and estuaries.

28.   Little Cormorant *Phalacrocorax niger*

Identification: 51 cm. The smallest cormorant. Shorter beak, thicker head and shorter neck separate from the other cormorants. **Adult breeding**: Glossy black. Upper back and wing-coverts silvery-grey scalloped with black. Short crest. White feathers on the head. **Adult non-breeding**: No crest. Black with white throat. **Immature**: Brown with paler scalloping on back. Paler below with throat and centre of abdomen white. (*See* Indian Cormorant, 27)
Range: Resident with local movements. Entire Southwestern India.
Habitat: Lakes, reservoirs, tanks, rivers, hill-streams and estuaries.

29.   Oriental Darter *Anhinga melanogaster* (Indian Darter)

Identification: 90 cm. A black long-necked cormorant readily identified by the long dagger-like beak. Tail long and stiff. Swims with snake-like head and neck showing out of water. **Adult**: Black. Back and wings streaked and speckled with silvery-grey. Head and neck chocolate-brown. White on chin, throat and along the sides of the neck from behind the eyes. Black below. **Immature**: Overall brown with paler neck. Streaked (duller) on back.
Range: Resident with local movements. Entire Southwestern India.
Habitat: Lakes, reservoirs, rivers and hill-streams.

Family FREGATIDAE

Large. Black and white seabirds. Wings long, slender and pointed. Tail deeply forked. Beak long and hooked. Soaring flight reminiscent of raptors. Throat bare and brightly coloured gular skin (males). Legs short and feathered with webbed feet.

31.   Great Frigate Bird *Fregata minor* (Lesser Frigate Bird)

Identification: 87–102 cm. A large soaring seabird with deeply forked tail. **Male**: Glossy black with a brown band on wings. **Female**: Larger than male. Black with paler throat. White breast and sides diagnostic.
Range: Stray records from coasts of Mumbai, Karnataka and Kerala.
Habitat: Marine. Occasionally closer to the coast.

Family ARDEIDAE

Long-legged wading birds. Neck long and slender; retracted in the form of an 'S' in flight. Sharp, dagger-like beak. Sexes generally alike.

36.   Grey Heron *Ardea cinerea*

Identification: 98 cm. A tall pale-looking marsh bird with long S-shaped neck. Solitary or in loose flocks. **Adult**: Ashy-grey with white crown and neck. Long, occipital crest black. Whitish below. Black-dotted line along sides of foreneck. In flight mostly greyish underside of wings diagnostic. **Immature**: Brownish tinge on plumes.
Range: Resident. Entire Southwestern India.
Habitat: Tanks, lakes, reservoirs, estuaries, mangrove swamps, salt pans and rocky shores.

37.   Purple Heron *Ardea purpurea*

Identification: 97 cm. A tall, dark and solitary long-necked marsh bird separated from grey heron by a preference for hiding among aquatic vegetation and a more pronounced bulge on neck in flight. **Adult**: Purplish blue or slaty above; blackish on wings and tail. Crown slaty-black. Neck rufous with black streaks. Chin and throat white. Slaty-black and chestnut on underside. In flight dark ferruginous underwing diagnostic. **Immature**: Browner. Sandy-rufous on head and neck.
Range: Resident with local movements. Entire Southwestern India.
Habitat: Tanks, lakes and rivers, especially those where there is a lot of floating and bordering vegetation such as lotus and reeds.

38.   Striated Heron *Butorides striatus* (Little Green Heron)

Identification: 46 cm. A medium-sized green-grey hunchbacked marsh bird. Solitary along edge of water and in overhanging vegetation. **Adult**: Forehead, crown and long occipital crest glossy greenish-black. Head and sides of neck grey. Back and wings mostly glossy green and grey. Grey and white below. **Immature**: Browner with heavy streaks on underside.
Call: .. *Keyow..or..Keyak*. Also rendered as *tewn-tewn-tewn* resembling redshank.
Range: Resident. Entire Southwestern India.
Habitat: Lakes, tanks, reservoirs, rivers and hill-streams with overhanging vegetation such as *Pandanus,* estuaries and mangrove swamps.

42.   Indian Pond Heron *Ardeola grayii*

Identification: 46 cm. A medium-sized sluggish earthy-brown marsh bird readily identified by its reluctance to fly on close approach. In flight, contrasting white wings and brown/maroon back diagnostic. **Adult breeding**: Head and neck yellowish-brown with a long occipital crest. Back deep maroon. Belly, wings and tail pure white. Occasionally with bright pink legs. **Adult non-breeding:** Brown with paler streaks on neck and breast. Grey-brown on back. Wings and tail white.

Range: Resident with local movements. Entire Southwestern India up to 2000 m in Nilgiris.

Habitat: All kinds of wetlands including paddyfields, rivers, hill-streams, estuaries, mangrove swamps, salt pans, beaches, city drains, garbage dumps, drier cultivation, coconut gardens and urban meadows.

44. Cattle Egret *Bubulcus ibis*

Identification: 51 cm. A medium-sized long-legged white bird following cattle. Comparatively shorter neck and legs separate from other white egrets at rest and in flight. **Adult breeding:** Golden-buff head, neck and back. White body, wings and tail. **Adult non-breeding:** White. Black legs and feet. Bright yellow beak diagnostic.

Range: Resident with local movements. Entire Southwestern India up to 2100 m in the hills.

Habitat: Wetlands, estuaries, meadows, cultivated areas, urban yards and open forests.

46. Great Egret *Casmerodius albus* (Large Egret)

Identification: 90 cm. A tall white marsh bird with a kink on its long neck. Long ornamental plumes absent on breast. Solitary. **Adult breeding:** White. Beak black (sometimes yellow at base). Legs and feet black with traces of pink-red. Ornamental plumes on back only. **Adult non-breeding:** Yellow beak (black-brown tip). (*See* Intermediate Egret, 47)

Range: Resident with local movements. Widespread in Southwestern India.

Habitat: Large tanks, lakes and reservoirs, rivers, estuaries and salt pans.

47. Intermediate Egret *Mesophoyx intermedia* (Median, Smaller Egret)

Identification: 80 cm. A white long-necked marsh bird intermediate in size between larger egret and little egret. S-shaped neck without a kink separates from large egret. Flocks. **Adult breeding:** White. Beak black (yellow at base). Ornamental filamentous plumes both on back and breast. Legs and feet black. **Adult non-breeding:** Yellow beak (dusky tip).

Range: Resident with local movements. Entire Southwestern India.

Habitat: All kinds of wetlands, estuaries and mangrove swamps.

49. Little Egret *Egretta garzetta*

Identification: 63 cm. A medium-sized white egret identified by its smaller size and black beak. Legs black with yellow feet. Gregarious. **Adult breeding:** Filamentous plumes on back and breast. Occipital crest. **Adult non-**

breeding: Crest and filamentous plumes absent. (*See* Indian Reef Heron, 50)

Range: Resident with local movements. Entire Southwestern India up to at least 900 m.

Habitat: Wetlands including paddyfields, estuaries, mangrove swamps and salt pans.

50. Western Reef Egret *Egretta gularis* (Indian Reef Heron)

Identification: 63 cm. A medium-sized blue-grey egret with yellow beak. White colour phase similar to little egret. Solitary. Coastal. **Adult breeding**: (1) White. Black legs with yellow feet. Indistinguishable from little egret except for the beak which may be horny brown or yellow. (2) Slaty-grey or slaty blue-black. White throat and neck. Yellow beak. **Immature**: Paler grey with whitish underparts.

Range: Resident with local movements. Entire West Coast.

Habitat: Sandy and rocky beaches, estuaries, mangrove, salt pans and occasionally inland rivers and lakes.

52. Blackcrowned Night Heron *Nycticorax nycticorax* (Night Heron)

Identification: 58 cm. A stocky grey, black and white marsh bird with a hunched appearance at rest. Beak prominent. Crepuscular. In flight flocks calling .. *wock* .. diagnostic. Large roosts in urban groves. **Adult**: Grey above with glossy black back and shoulders. Crown and occipital crest black. Underparts largely white. **Immature**: Grey-brown, streaked and spotted with buff.

Call: ..*Wock*..or.. *Kwaark* ... (in flight).

Range: Resident with local movements. Entire Southwestern India.

Habitat: All wetlands including estuaries and mangrove swamps.

53. Malayan Night Heron *Gorsachius melanolophus* (Malay/Tiger Bittern)

Identification: 51 cm. A stocky rufous, black and chestnut heron with bold streaks. In flight dark primaries with buff tips diagnostic. Forests. **Adult**: Crown and long drooping crest slaty-black. Chin and throat white. Rest of plumage mostly chestnut-cinnamon and white (below) with black streaks and barring. Flight feathers and tail black with paler tips. **Immature**: Black head. Dark brown streak down white throat. Rest of plumage brown and buff spotted with white and brown. (*See* Black Bittern, 58 and Great Bittern, 59)

Range: Resident with local movements. Range in Southwestern India discontinuous. Locally known from Kerala, Nilgiris and western Karnataka up to Belgaum–Goa.

Habitat: Marshes and streams within dense evergreen or semi-evergreen forest.

56.   Cinnamon Bittern *Ixobrychus cinnamomeus* (Chestnut Bittern)

Identification: 38 cm. A small rufous-cinnamon marsh bird with a black line down foreneck. In flight, pale underwing diagnostic. Solitary. **Adult Male**: Chestnut-cinnamon or rufous above. Paler below. **Adult Female**: Black crown. Wings and shoulders spotted with buff. Underparts mostly streaked with brown. **Immature**: Like female; heavier spots and streaks.
Call: Rendered as .. *kok-kok* .. or .. *ek-ek-ek* .. or .. *gook-gook* .. *gook* ...
Range: Resident with local movements. Entire Western Southwestern India up to at least 900 m.
Habitat: Reed-covered tanks and lakes, rice fields and rarely in estuaries.

57.   Yellow Bittern *Ixobrychus sinensis*

Identification: 38 cm. A small pale-coloured heron identified in flight by the contrasting dark wings. Solitary within dense reeds. Slow movements diagnostic. **Adult**: Crown and crest black. Sides of head pinkish. Largely straw-yellow on back. Wings and tail blackish. Pale yellowish-buff below. Blackish with streaks on breast. **Immature**: Rufous-brown with heavy streaking. Buff line down centre of foreneck prominent.
Call: Rendered as .. *kaka-kakak* ...
Range: Resident with local movements. Entire Southwestern India up to at least 1200 m.
Habitat: Reed covered tanks and lakes, rice fields and mangrove swamps.

58.   Black bittern *Ixobrychus flavicollis*

Identification: 58 cm. A medium-sized black marsh bird with buff chin and throat. Solitary. **Adult Male**: Black. Sides of neck yellowish. Throat white with a rufous dotted line down the middle. **Adult Female**: Brownish with paler streaks on underparts. **Immature**: Like female; scaly-streaked pattern on entire plumage. (*See* Malayan Night Heron, 53)
Range: Resident with local movements. Widespread in Southwestern India.
Habitat: Reedy marshes, hill-streams with vegetation overhanging and mangrove swamps.

59.   Great Bittern *Botaurus stellaris* (Bittern)

Identification: 71 cm. A large stocky marsh bird identified by the mottled buff, black and brown colour pattern. Solitary. **Adult**: Crown to upper back (including drooping crest) black. Throat and chin white with black median stripe. Rest of the plumage largely buff with darker black streaks. (*See* Malayan Night Heron, 53)
Range: Winter visitor. Stragglers known from Gujarat, Maharashtra, Karnataka and Tamil Nadu.
Habitat: Reed covered tanks and lakes.

Family CICONIIDAE

Large marsh birds with long legs and beak. Fly with legs and neck stretched out. Soaring flight reminiscent of raptors.

60. Painted Stork *Mycteria leucocephala*

Identification: 100 cm. A tall white, black and pink marsh bird with large yellow beak. Black flight feathers and breast band contrasting with white plumage diagnostic in flight. **Adult**: White with large yellow beak. Back with a bold scaly pattern of glossy black extending across breast. Rose-pink lower back and above tail. Tail black. **Immature**: Pale brown with scaly pattern on neck. No band on breast.
Range: Resident with local movements. Widespread along the drier parts of Southwestern India.
Habitat: Large tanks, lakes, marshes and rivers.

61. Asian Openbill Stork *Anastomus oscitans* (Openbill Stork)

Identification: 80 cm. The smallest of our storks. Grey and black plumage and open space in beak diagnostic. Gregarious. **Adult**: White (breeding) or grey with glossy black shoulders, wings and tail. **Immature**: Darker grey-brown above. Beak without the open space.
Range: Resident with local movements. Widespread in Southwestern India up to at least 600 m.
Habitat: Large tanks, lakes, wet cultivation and estuarine marshes.

62. Woollynecked Stork *Ciconia episcopus* (Whitenecked Stork)

Identification: 90 cm. A large black and white stork with red legs. Soars above forests. **Adult**: Glossy black with white neck and under tail-coverts. Naked face and beak black. **Immature**: Duller. (*See* Lesser Adjutant, 68)
Range: Resident. Widespread over entire Southwestern India up to at least 700 m.
Habitat: Wet fields, tanks, lakes, rivers and marshes within forests. Rarely, estuaries.

65. Black Stork *Ciconia nigra*

Identification: 95 cm. A large black stork with white underparts. Bright red beak and legs diagnostic. **Adult**: Black. Lower breast, belly and under tail-coverts white. **Immature**: Browner. Beak and legs yellowish.
Range: Rare winter visitor recorded in recent years in Uttara Kannada (Karnataka) and Periyar lake in Kerala.
Habitat: Salt pans and lakes. Occasionally soar over open forest.

68. Lesser Adjutant *Leptoptilos javanicus* (Lesser Adjutant Stork)

Identification: 113 cm. A very large black and white stork identified by its massive beak, naked head and neck. **Adult**: Glossy black. White underparts. Naked head and neck orange. Large yellowish beak. **Immature**: Brownish with more feathers on head and neck. (*See* Wodlynecked Stork, 62)

Range: Resident with local movements. Entire Southwestern India up to at least 500 m.

Habitat: Large tanks, swamps, flooded land and also open, dry cultivation.

Family THRESKIORNITHIDAE

Large, long-legged marsh birds with long curved beaks (ibises) or flattened spatulate beaks (spoonbills).

69. Blackheaded Ibis *Threskiornis melanocephalus* (White Ibis)

Identification: 75 cm. A large white stork-like bird with black neck, head and long curved beak. Red armpits visible in flight. **Adult breeding**: Greyish wings. Long ornamental plumes hanging from base of neck. **Adult non-breeding**: Entirely white plumage. Head and neck bare. **Immature**: Head and neck mostly covered with down. Bare patches under the wings not red.

Range: Resident with local movements. Widespread in Southwestern India up to about 1000 m.

Habitat: Rivers, reservoirs, marshes, flooded fields and estuaries.

70. Rednaped Ibis *Pseudibis papillosa* (Black Ibis)

Identification: 68 cm. A medium-sized glossy black stork-like bird with long down-curved beak. White patch on shoulders diagnostic. Red legs. **Adult**: Naked black head covered with red warts. **Immature**: Dull brown. Crown, head and throat feathered. (*See* Glossy Ibis, 71)

Range: Resident with local movements. Widespread along the drier eastern parts of Southwestern India.

Habitat: River banks, harvested fields, marshes and edges of tanks.

71. Glossy Ibis *Plegadis falcinellus*

Identification: 52 cm. A small glossy green-blue or blackish ibis with feathered head. **Adult non-breeding**: Head and neck brown with white streaks. Glossy green-blue shoulders and wings. **Immature**: Like non-breeding adult but duller with an ashy tinge on plumage. **Adult breeding**:

Rich chestnut or maroon. Head, neck, lower back, rump and tail glossy green-purple. Deep purple under tail.
Range: Winter stragglers along the eastern parts of Southwestern India.
Habitat: Marshes, lake and river banks.

72. Eurasian Spoonbill *Platalea leucorodia* (Spoonbill)

Identification: 60 cm. A large white stork-like bird with spatula-shaped black and yellow beak. **Adult breeding**: White bushy crest. Yellowish patch on foreneck. Naked yellow throat. Legs black. **Immature**: Traces of black on wings.
Range: Resident with local movements. Widespread in Southwestern India up to at least 600 m.
Habitat: Marshes, reservoirs, rivers and estuaries.

Family PHOENICOPTERIDAE

Tall, excessively long-legged marsh birds with long necks and peculiarly down-curved thick beaks. Pink, crimson, white and black plumage. Feet webbed.

73. Greater Flamingo *Phoenicopterus ruber*

Identification: 140 cm. A tall marsh bird identified by its white, rosy pink, crimson and black plumage. Stout down-curved red beak diagnostic. In flight outstretched long neck and legs and black wings characteristic.
Immature: Grey-brown. Pale pink underwing.
Range: Sporadic visitor to Southwestern India.
Habitat: Lakes, estuaries, tidal mudflats and salt pans.

Family ANATIDAE

Water birds of diverse sizes and colours. Flat beak. Slender neck. Narrow, pointed wings. Short tail. Legs short with webbed feet. Distinct male, female, breeding and non-breeding plumages.

NOTE: Sibley and Monroe (1990) treat the Whistling Teals as belonging to family DENDROCYGNIDAE.

88. Lesser Whistling Teal *Dendrocygna javanica*

Identification: 42 cm. A small brown and maroon duck with paler buff coloured neck. Feeble flight. Noisy whistling call. Gregarious. **Adult**: Brown. In flight chestnut-maroon wings with black flight feathers diagnostic.

Uniform chestnut upper tail-coverts (vs. creamy-white) separate from large whistling teal. **Immature:** Duller brown. (*See* Fulvous Whistling Teal, 89) **Call:** Shrill whistles in flight. Rendered as .. *seasick..* **Range:** Resident with local movements. Entire Southwestern India. **Habitat:** Marshes and lakes with floating vegetation, larger reservoirs and open sea.

89.   Fulvous Whistling Teal *Dendrocygna bicolor* (Large Whistling Teal)

Identification: 51 cm. A medium-sized brown duck rather similar to lesser whistling teal. Larger size, cream coloured upper tail-coverts, faster flight and smaller flocks diagnostic. **Adult:** Head and neck pale buff without a darker cap. Prominent black line down hindneck. Broad, diffuse whitish collar around middle foreneck. **Immature:** Duller brown.
Call: Described as similar to that of lesser whistling teal.
Range: Resident with local movements. Widespread in Southwestern India.
Habitat: Reed-covered tanks and lakes bordered with tall vegetation.

90.   Ruddy Shelduck *Tadorna ferruginea* (Brahminy duck)

Identification: 66 cm. A large goose-like orange-cinnamon-brown duck with pale creamy head and neck. Contrasting black primaries diagnostic in flight. **Adult Male:** White and orange-brown. Metallic green wing speculum. Black beak, legs and tail. Narrow black collar around base of neck (breeding). **Adult Female:** Lacks the collar. Paler head. **Immature:** Similar to female with more greyish tinge on wings.
Call: A nasal goose-like .. *aang* .. *aang* ...
Range: Winter visitor. Coastal northern Karnataka.
Habitat: Estuaries and salt pans.

91.   Common Shelduck *Tadorna tadorna*

Identification: 61 cm. A large, brightly coloured duck identified by its black, white and chestnut plumage, red upturned beak and pink legs. Gregarious. **Adult Male:** Head and neck glossy greenish-black. Rest of the upperparts white with two broad black bands along the upper back. Metallic green wing speculum bordered above by chestnut. Chestnut girdle on upper breast and shoulders. Black band from breast to vent. Conspicuous red knob on beak. **Adult Female:** Smaller and duller. Breast band with black vermiculations. No knob on beak. **Immature:** Plumage on head and the upperparts browner. Face and throat whitish. No chestnut breast band. Pink beak and grey legs.
Range: Winter visitor. Northern parts of Southwestern India (Gujarat and Maharashtra).
Habitat: Lakes and large reservoirs.

93. Northern Pintail *Anas acuta* (Pintail)

Identification: 56 cm (male 74 cm including tail). A medium-sized, slender duck with long neck and pointed tail. Peculiar trident ending with legs on either side of pointed tail diagnostic in flight. Flocks in thousands. **Male breeding**: Chocolate-brown head, face and throat. Black hindneck. White band along sides of neck running broader on breast and belly. Upper plumage mostly grey. Long tail and under tail-coverts black. Lower belly with a buff patch. Green wing speculum. **Female**: Mottled brown and buff with pointed but pinless tail. Trailing white edge to inner half of brown wing. Brown (indistinct) wing speculum (vs. purple-blue in female mallard). **Non-breeding** and **Immature Males**: Like female. Traces of ashy grey-white on back and wings.
Call: Very similar to that of domestic duck. .. *quack* ...
Range: Winter visitor. Locally abundant over entire Southwestern India.
Habitat: Lakes, reservoirs, estuaries and salt pans.

94. Common Teal *Anas crecca*

Identification: 38 cm. A small, mottled brown-grey gregarious duck. Green wing speculum with whitish border diagnostic in flight. **Male breeding**: Greyish with chestnut head. Broad green band running between eyes and nape narrowly bordered with white. **Female**: Mottled brown. Darker wing-coverts. Whitish belly. Prominent green wing speculum (vs. less striking in female garganey). **Non-breeding Male** and **Immature**: Head like female with traces of blackish-brown. Scaly pattern on back with rufous edging. Belly spotted (immature). (*See* Garganey, 104)
Call: Rendered as a shrill .. *krit* .. *krit* .. (male) and a harsh .. *gueck* .. (female).
Range: Winter visitor. Entire Southwestern India.
Habitat: Lakes, reservoirs, tanks and salt marshes.

97. Spotbilled Duck *Anas poecilorhyncha* (Spotbill Duck)

Identification: 61 cm. A large blackish duck with paler head and neck. Bright colour pattern on beak diagnostic. Pairs or small flocks. **Adult**: Dark brown and grey with a scaly effect on upperparts. Metallic green wing speculum bordered above with white (conspicuous in flight). Bright red legs. Yellow-tipped blackish beak with two red knobs at base. **Immature**: Paler than adult. No red on beak.
Call: Similar to that of domestic duck .. *quack* ... Male shriller.
Range: Resident with local movements. Widespread in Southwestern India. Scarce south of Karnataka.
Habitat: Large tanks, reservoirs and lakes, often surrounded by forests. Rarely rivers.

100. Mallard *Anas platyrhynchos*

Identification: 61 cm. A large duck resembling the domestic duck with olive-yellow-green beak and orange legs. In flight white-bordered purplish wing speculum. Rounded greyish tail. Small flocks. **Male breeding**: Grey above. Metallic green head and neck separated from chestnut breast by white collar. Black on rump and upper tail-coverts including the up-curled feathers. **Female, non-breeding Male** and **Immature**: Generally similar. Brown, buff with streaks and spots of black. Dark line through eye. Orange legs and purple-blue wing speculum separate from all similar ducks. (*See* Northern Pintail, 93 and Spotbilled Duck, 97)
Call: Like that of domestic duck. Rendered as a single loud .. *quack* .. or .. *quack* .. *quack* .. *quack* ... Male shriller.
Range: Winter visitor. Northern Maharashtra.
Habitat: Shallow tanks with floating vegetation and marshes.

101. Gadwall *Anas strepera*

Identification: 50 cm. A medium-sized duck with rounded tail. In flight, conspicuous white speculum on wing diagnostic. Grey beak and yellow-orange legs. Gregarious in tanks with floating vegetation. **Male breeding**: Greyish-brown with dark brown head. White belly. Black tail. Chestnut, black and white pattern on wing diagnostic even at rest. **Female, non-breeding Male** and **Immature**: Spotted grey-brown. Wing speculum conspicuous only in flight. Smaller size, white belly and darker tail separate from female mallard. (*See* Northern Pintail, 93)
Call: Described as similar to mallard's but softer.
Range: Winter visitor. Entire Southwestern India. Rarer south of Karnataka.
Habitat: Tanks and marshes with floating vegetation. Rarely open water in reservoirs.

103. Eurasian Wigeon *Anas penelope* (Wigeon)

Identification: 49 cm. A medium-sized stumpy duck with short beak and rounded head. Blue-grey beak and legs. Gregarious. **Male breeding**: Grey. Chestnut or rusty-red head with creamy patch on crown. Pinkish breast. White bar on folded wings. Black tail-coverts. In flight white shoulder patch diagnostic. **Female** and **non-breeding Male**: Reddish-brown above with black vermiculations. White below. **Female**: separated from other similar female ducks by more rufous, less mottling and whiter belly. No dark streak through eye. No pale supercilium. In flight, paler grey wings contrast with dark flight feathers.
Call: Rendered as .. *whee-oo* ... (male) and a short .. *quack* ... (female).
Range: Winter visitor. Rare in the south.
Habitat: Reed-covered marshes. Rarely in salt marshes.

104. Garganey *Anas querquedula* (Bluewinged Teal)

Identification: 40 cm. A small grey fast-flying duck. Grey-blue shoulders conspicuous in flight. Flocks in thousands. **Male breeding**: Grey-brown with paler and black markings. Rosy-brown head and neck with a prominent white eyebrow. Long, black and white scapulars. Metallic green wing speculum bordered by white visible in flight. **Female, non-breeding Male** and **Immature**: Head brown with conspicuous white supercilium. Throat white. A dark line through eyes. Shoulders blue in non-breeding male; grey-brown in female. Green wing speculum. White belly. Female separated from similar female common teal by white throat (vs. mottled) and in flight by grey shoulders and unglossed blackish wing speculum.
Range: Winter visitor. Entire Southwestern India.
Habitat: Marshes, reservoirs, lakes, estuaries and salt pans.

105. Northern Shoveller *Anas clypeata* (Shoveller)

Identification: 51 cm. A medium-sized duck with a large shovel-like blackish beak. Legs orange. In flight, blue-grey and black wings diagnostic. Small flocks. **Male breeding**: Glossy green head and neck. Shoulders grey with a white band bordering the green wing speculum. Breast and sides of flanks white. Rest of the underparts mostly reddish-chestnut. **Female and nonbreeding Male**: Mottled brown with a dark stripe through eyes. Bright wing pattern identifies non-breeding male. Female has orange base of beak. **Immature**: Lacks the mottling on upperparts.
Call: Described as similar to mallard's but softer.
Range: Winter visitor. Widespread in eastern Southwestern India.
Habitat: Large lakes, reservoirs and tanks.

107. Redcrested Pochard *Netta rufina*

Identification: 54 cm. A medium-sized, crested duck identified by its orange head and red beak. Underparts black with white flanks. In flight white underside of wings and a broad white bar along trailing edge of wings diagnostic. **Male breeding**: Upperparts pale brown. White shoulders and wing speculum. **Female, non-breeding Male** and **Immature**: Dull sooty-brown with dark brown crown and nape. Underparts largely whitish. Female identified by its blackish beak and lack of fluffy crest. (*See* Common Pochard, 108 and Ferruginous, 109).
Range: Winter visitor. Widespread north of Karnataka.
Habitat: Large tanks with floating vegetation and reservoirs.

108. Common Pochard *Aythya ferina*

Identification: 48 cm. A small, squat duck with black and blue beak. High crown with sloping forehead diagnostic. In flight, distinguished from other

pochards by absence of white wingbar. Large flocks. **Male breeding**: Head
and neck chestnut-red. Breast, upper back and tail black. Rest of the
plumage pale grey (whitish). **Male non-breeding**: Head duller. Breast and
upper back brown. **Female** and **Immature**: Like non-breeding male but
with touches of grey in the plumage. Face and throat of female paler buff.
Browner underparts and smaller size separate from female redcrested
pochard. (*See* Ferruginous Pochard, 109)
Range: Winter visitor. Southwestern India as far south as Karnataka.
Habitat: Tanks with submerged vegetation and reservoirs.

109. Ferruginous Pochard *Aythya nyroca* (White-eyed Pochard)

Identification: 41 cm. The smallest pochard. Rufous-blackish-brown with
white eyes. In flight white oval patch on belly, white wing speculum and
under tail-coverts diagnostic. Frequently in sea. **Male breeding**: Rich
brown and white. **Male non-breeding**: Reddish-brown and white. **Female**
and **Immature**: Like male but duller with less contrasting belly patch.
Brown eyes. Fully brown head and neck separate from other female
pochards.
Range: Winter visitor. Southwestern India as far south as Kerala.
Habitat: Reservoirs, estuaries and sea.

111. Tufted Duck *Aythya fuligula*

Identification: 43 cm. A small crested black and white duck. Grey beak and
high forehead diagnostic. In flight white band on wings conspicuous. **Male
breeding**: Jet-black. White abdomen, flanks and wing speculum. A tuft of
feathers hanging down from the nape. Yellow eyes. **Non-breeding Male**,
Female and **Immature**: Brownish with the white portions less distinct.
White mottling on plumage. Female identified by shorter tuft and a white
patch on base of beak. Pale eyes diagnostic. (*See* Ferruginous Pochard,
109 and Great Scaup Duck, 112)
Call: Rendered as a harsh .. *kurr-r-r* ... in flight.
Range: Winter visitor. Southwestern India as far south as Karnataka along
the eastern parts.
Habitat: Open deep waters in tanks and reservoirs.

112. Greater Scaup *Aythya marila* (Scaup Duck)

Identification: 46 cm. A small black and white duck very similar to tufted
duck. No occipital tuft. Beak grey. Broader white band on wing diagnostic
in flight. **Male breeding**: Glossy black head, neck and tail. White underparts. Back pale grey with vermiculations. **Non-breeding Male**, **Female**
and **Immature**: Dark brown and whitish. Female identified by a broad

white band on forehead around base of beak; more prominent than in female tufted duck.
Range: Winter straggler recorded from northern Maharashtra.
Habitat: Freshwater lakes.

114. Cotton Pygmy Goose *Nettapus coromandelianus* (Cotton Teal)

Identification: 33 cm. The smallest of our ducks. Greenish-grey and white with short beak. Gregarious. **Male breeding**: Head, neck and underparts white. Crown, collar and back glossy green-blackish. White trailing edge of wings diagnostic in flight. **Non-breeding Male, Female** and **Immature**: Duller with less striking colour contrast. Dark line through eyes and a less striking white bar on wings identifies the female.
Call: Rendered as .. *fix baggonets* .. in flight.
Range: Resident with local movements. Entire southwestern India.
Habitat: Lakes, reservoirs and tanks with reeds and floating vegetation.

115. Comb Duck *Sarkidiornis melanotos*

Identification: 76 cm. A very large glossy black and white duck; a giant version of the cotton pygmy goose. Pairs or small flocks. **Male**: White with black spots on face and neck. Fleshy comb at base of beak diagnostic. Glossy purplish-green upper back and greyish lower back. Black bands down sides of breast and across under tail-coverts. In flight bronzy wing speculum and greyish back diagnostic. **Female** and **Immature**: Lack fleshy comb on beak. Duller and more mottled below. The black bands on breast and under tail-coverts absent.
Call: Harsh croaks rendered as .. *honk* ...
Range: Resident (?) with local movements. Mostly recorded as a visitor as far south as northern Karnataka.
Habitat: Larger lakes, reservoirs and shallow tanks with floating vegetation.

Family ACCIPITRIDAE

Predatory birds including kites, hawks, buzzards, hawk-eagles, eagles, ospreys, vultures and harriers. Generally a group of confusingly similar species in different colour phases and juvenile–adult plumages. Most species soar and the broad groups can be separated by body size and shape of wings or tail. Sexes generally alike except in size, females being larger. **Kites** and **Harriers**: large and slender with long narrow wings and tails. Long, bare legs. Graceful flight. Harriers differ from kites in their rounded owl-like faces, unforked tails and a greater preference for open grasslands and cultivation. All harriers are winter visitors to the Western Ghats. **Bazas**: small or medium-sized forest birds of prey with long, erect crests. **Hawks**:

small birds of prey with short rounded wings, long tails and long, bare legs. Rapid wing beats in flight. **Buzzards**: large and stocky hawks with broader and longer wings. All true buzzards *(Buteo)* are winter visitors. **Eagles**: large with more prominent beaks. Long wings with primary feathers splayed like fingers in flight. Short squared tails. Feathered legs. **Hawk-Eagles**: large hawks. Longer tail than eagles and often a prominent crest. Legs feathered. **Vultures**: very large and gregarious. Unfeathered legs, head and neck. Soaring flight on long wings held like a 'V'. **Ospreys**: large whitish and black fishing hawks. Vicinity of water.

NOTE: Some authorities treat the Osprey as belonging to family PANDIONIDAE.

124. Blackwinged Kite *Elanus caeruleus*

Identification: 33 cm. A small grey and white bird of prey with a black patch on shoulder diagnostic both at rest and in flight. Short tail. Flight gull-like. Often hovers. Solitary. **Adult**: Pale grey above. White below. Blood-red eyes. Black tips of closed wings extend beyond tail at rest. **Immature**: Brownish plumage with paler streakings and scalloping. (*See* Pallid Harrier, 190)

Call: Described as an occasional high-pitched squeal or whistle.

Range: Resident. Entire Southwestern India up to at least 1200 m.

Habitat: Open forests, high altitude grasslands and edges of montane forests, cultivation, open scrub and urban wastelands.

126. Jerdon's Baza *Aviceda jerdoni* (Legge's Baza)

Identification: 48 cm. A medium-sized brown bird of prey with a prominent erect crest. At rest, wing tips almost reach tip of tail. Flight, buzzard-like. Pairs or small flocks. **Adult**: Brown with rufous and black head. Broad white tips to black crest feathers. Rufous-brown underparts (whitish in females) with broad rufous bands. Tail brown with three broad bands; the terminal one being broadest and darkest. **Immature**: Similar to female. Tail with 4–5 broad bands. Barring on underparts feebler. General appearance similar and confusable with crested goshawk. However longer crest (at rest), longer wings with broader and closer banding on primaries (in flight) and upper mandible toothed near tip (in hand) separate the baza from the crested goshawk.

Call: Calls rendered as .. *kip-kip-kip* .. (similar to that of palm squirrel) or .. *kikaya* .. *kikaya* .. during courtship; .. *pee-ow* .. in flight and a long drawn .. *gueeer* .. confusable with that of crested goshawk.

Range: Resident. Western Ghats southwards from Coorg (Karnataka) and Wynaad (Kerala) up to 900 m.

Habitat: Evergreen forests and associated habitats.

127. Black Baza *Aviceda leuphotes* (Blackcrested Baza)

Identification: 33 cm. A small, black and white bird of prey with a long erect crest. Broad wings with black lining, black under tail-coverts contrasting with paler tail and belly and white band on dorsal side of wings diagnostic. Flight crow-like. Flocks. **Adult**: Black above. White blotches on wings. Throat, breast and under tail-coverts black. Rest of the underparts white with broad black, maroon and rufous bands.

Call: Described as a plaintive whistle similar to the mewing of a gull.

Range: Resident. Western Ghats, southwards from Nilgiri–Wynaad (Kerala) along the foothills.

Habitat: Evergreen forests, associated habitats and along streams.

130. Oriental Honey Buzzard *Pernis ptilorhynchus*
(Crested Honey Buzzard)

Identification: 68 cm. A large hawk with a short crest. Slim appearance with small head on longish neck resembling that of a pigeon or cuckoo diagnostic in flight. Wings held flat and in line with the shoulders while soaring. Underside of wing silvery-grey with dark banding. Tail long and narrow. **Adult**: Colour variable: pale whitish-grey-brown to almost black. Normally grey above with darker head and crest. Wings and tail closely barred. Pale brown below barred with white. Barring on underparts may be totally absent in paler forms. Legs unfeathered. Scale-like feathers on lores and around eyes diagnostic in hand.

Call: A long drawn .. *wheeew* .. sounding like a scout whistle. Heard during early night as well.

Range: Resident with local movements. Western Ghats up to at least 1000 m.

Habitat: Open deciduous and evergreen forests and associated secondary habitats such as monocultures and cultivation.

133, 134. Black Kite *Milvus migrans* (Pariah Kite)

Identification: 61–66 cm. A familiar dark, large and slender bird of prey with long, deeply forked tail. In flight, slow deliberate wing beats. **Adult**: Dark brown. Pale patch on underside of wing in flight. Fork on tail not clear in moulting birds. **Immature**: Paler with more speckles and streaks on head. Race *lineatus* is larger than race *govinda*. White patch on underwing more conspicuous in flight.

NOTE: Some authorities consider race *lineatus* as a full species.

Call: A shrill whistle .. *eehee-he-he-he-he-he-he* .. reminiscent of the neighing of a horse. Also rendered as .. *ewe-wir-r-r-r-r* .. at rest and in flight. Sharp shrieks and squeals often uttered during courtship and aggression.

Range: Resident. Entire Southwestern India. Race *lineatus* is a winter visitor through Maharashtra and south up to Karnataka.

Habitat: Mostly urban though frequently seen in cultivation, marshes, estuaries and beaches. Race *lineatus* prefers open forests.

135. Brahminy Kite *Haliastur indus*

Identification: 48 cm. A familiar medium-sized rusty-red bird of prey with white head, neck, upper back and breast. Short rounded tail. Near water. **Adult**: White feathers streaked finely with black. In flight pale chestnut wings with black tips diagnostic. **Immature**: Brown like black kite. Tail shorter and rounded. Paler patch on underwing confusable with buzzards (*Buteo*) from which separated by lack of dark banding on flight feathers, less rounded wings and uniformly dark tail.

Call: Short squeal. More nasal than that of black kite. Sometimes .. *tehee-tee-tee-tee-tee-tee* ...

Range: Resident. Entire Southwestern India.

Habitat: Urban with a greater fondness for wetter habitats such as paddyfields, marshes, rivers, estuaries, mangroves, beaches and often out over open sea following trawling boats.

136. Northern Goshawk *Accipiter gentilis* (Goshawk)

Identification: 50–61 cm. A large hawk resembling shikra. In flight, broad rounded wings, closely barred underparts and longish tail with 3–4 broad bands diagnostic. **Adult**: Dark grey above with a white supercilium. Underparts white cross-barred with black. **Immature**: Brown upperparts with white streakings; broader on nape and hindneck. Tail with 4–5 broad bands. Underparts buff with bold blackish drops or spots (vs. bars).

Call: Described as a cackling *gaik .. gaik .. gaik* ...

Range: Winter straggler recorded from Gujarat and recently in Pune (Maharashtra).

Habitat: Open forests.

137, 138, 139. Shikra *Accipiter badius*

Identification: 30–36 cm. A small grey and whitish hawk. In flight whitish underwings and contrasting black primaries diagnostic. Red eyes. **Adult Male**: Ashy-blue-grey above. White below finely cross-barred with rusty-brown especially on breast. Faint grey mesial stripe on throat. White under tail-coverts conspicuous in flight. Tail appears unbanded (dorsally) at rest. **Adult Female**: Browner and larger than male. Central tail feathers with a distinct subterminal band; other bands very faint or absent. **Immature**: Brown above with a scaly pattern. Dark bands on tail and underwing. Dark mesial stripe on whitish throat. Broad brown bands and streaks on breast and belly.

The races are barely distinguishable in the field. Race *cenchroides* is larger and paler than races *dussumieri* and *badius* (with the barring on underparts extending to the thighs and vent). Race *badius* is the smallest. **Call**: Aggressive *tihee .. tihee .. tihee ...* also rendered as *.. eheeya .. eheeya ...* **Range**: Resident. Entire Southwestern India up to 1500 m. Races *dussumieri* and *badius* are separated in Kerala; the latter representing the species in the south. Race *cenchroides* is a winter straggler into Gujarat and Maharashtra.

Habitat: Open secondary evergreen and deciduous forests, scrub and cultivation, urban gardens and groves.

145. Crested Goshawk *Accipiter trivirgatus*

Identification: 31–36 cm (female larger). The only crested sparrow-hawk. Rather heavier built than shikra. Shorter crest separates from the similar Legge's baza. **Adult**: Dark grey-brown above darker on crest. Black mesial stripe on throat and a dark line on cheeks. White underparts with rusty-brown streaks on breast and barring on belly. Eyes yellow. White under tail-coverts conspicuous in flight. **Immature**: Browner with a scaly pattern on head and nape. Black mesial stripe on white throat. Pale supercilium. Underparts white with dark brown streaks and spots on breast and belly. (*See* Shikra, 137, 138, 139)
Call: A long drawn *.. gueeer ..* repeated mournfully from a perch. Confusable with that of Legge's baza and small yellownaped woodpecker.
Range: Resident. Western Ghats from Goa to southern Kerala up to 1000 m.
Habitat: Evergreen forests, moist secondary and deciduous forests, open grasslands in the higher hills and occasionally within urban limits of hill towns.

147. Eurasian Sparrow-Hawk *Accipiter nisus* (Asiatic Sparrow-Hawk)

Identification: 31–36 cm (female larger). A small grey and rufous sparrow-hawk slightly bigger than shikra. No mesial streak. Rufous cheeks diagnostic. Prefers denser vegetation than shikra. **Adult**: Dark grey-brown above. White below faintly barred with rufous on breast and abdomen. Tail appears banded at rest. Yellow eyes. White supercilium. Legs slenderer and longer than in shikra. Rufous cheek patch not clear in the browner female. **Immature**: Brown above with a scaly effect. White underparts with brown barring on breast and belly separates from all other immature sparrow-hawks.
Call: *.. ti .. ti .. tili .. li .. li .. li ...*
Range: Winter visitor. Entire Southwestern India up to 800 m.
Habitat: Evergreen and secondary forests and densely planted groves within urban limits.

151. Besra *Accipiter virgatus* (Besra Sparrow-Hawk)

Identification: 29–34 cm (female larger). The smallest sparrow-hawk on the Western Ghats. Darker than shikra. Striking black mesial stripe on white throat diagnostic. **Male**: Blackish-slaty-grey above. Tail grey with bands showing. White chin and throat. Breast and belly rufous with darker streaks and bands. White under tail-coverts. **Female**: Upperparts browner than male. White below with rufous-brown streaks and bands. Black mesial stripe. **Immature**: Not distinguishable from other juvenile sparrow-hawks in the field except by smaller size.

Call: Somewhat similar to that of Eurasian sparrow-hawk .. *tee* .. *tee* .. *tee* .. *tee* .. *tee* ... Also rendered as .. *tchew* .. *tchew* .. *tchew* ...

Range: Resident. Western Ghats up to 1200 m.

Habitat: Dense evergreen, moist deciduous and secondary forests.

153. Longlegged Buzzard *Buteo rufinus*

Identification: 61 cm. A small eagle with whitish head and contrasting rufous-brown belly resembling immature brahminy kite. Separated from the commoner crested honey buzzard by broader wings with tips splayed, broader tail and shorter neck. Tail shorter than that of hawk-eagle. **Adult**: Plumage very variable. Brown or reddish-brown to pale sandy with paler (sometimes almost white) head. Characteristic underwing pattern of white bordered by black on flight feathers diagnostic in overhead flight. Rufous tail almost unbanded. Unfeathered legs. **Immature**: Differs from adult in having banded tail. (*See* Common Buzzard, 156)

Range: Winter visitor. Rare. As far south as Karnataka along the foothills.

Habitat: Cultivation, open scrub country and deciduous forest.

156. Common Buzzard *Buteo buteo* (Buzzard)

Identification: 51–56 cm. A medium-sized rufous-brown buzzard smaller and shorter-winged than the similar longegged buzzard. **Adult**: Plumage variable. Darker head, less rufous on tail and a less distinct wing pattern possibly separate from the longlegged buzzard in field. Bare legs. Brown eyes. **Immature**: Tail more rufous with narrow barring. Lacks the broad subterminal band sometimes present in adults.

Range: Winter visitor. Rare. As far south as Kerala.

Habitat: Open country.

157. White-Eyed Buzzard *Butastur teesa* (White-Eyed Buzzard-Eagle)

Identification: 43 cm. A small, greyish-brown buzzard with white eyes. Whitish underside of broad, blunt wings contrasting with darker body diagnostic in flight. **Adult**: White throat with three dark stripes running down.

White patch on nape. Unfeathered legs. Closed wings reach almost to the tip of rufous tail at rest. Greyish shoulders conspicuous in flight. **Immature**: Whitish head. Buff underparts streaked with brown on breast and belly. Stripes on throat sometimes absent.
Call: Rendered as a mewing .. *pitweer .. pitweer ...*
Range: Resident. Entire Southwestern India up to at least 600 m.
Habitat: Dry cultivation, scrub and open grasslands along the hill slopes.

159. Mountain Hawk-Eagle *Spizaetus nipalensis* (Feathertoed Hawk-Eagle)

Identification: 70 cm. A largish, slender eagle with a long crest. Underparts brown. In flight pale grey appearance with short rounded and upturned wings diagnostic. **Adult**: Dark brown above barred with white on rump and upper tail-coverts. Tail dark brown barred with dark grey. Three black streaks down chin and throat. Breast whitish with black spots and streaks. Rest of underparts barred with chocolate-brown. Darker brown underparts help separate from the similar changeable hawk-eagle. In hand, feathers on tarsus covering base of middle toe diagnostic.
Call: Rendered as a loud scream and also .. *klu-weet-weet* .. or .. *kee-kikik* ...
Range: Resident. Rare. Western Ghats of Karnataka, Nilgiris, Palnis and Kerala up to 1200 m.
Habitat: Evergreen forest.

161. Changeable Hawk-Eagle *Spizaetus cirrhatus* (Crested Hawk-Eagle)

Identification: 72 cm. A large, slender crested bird of prey similar to feathertoed hawk-eagle. Whitish underparts and rufous-brown tail barred with black, diagnostic. Commoner than mountain hawk-eagle and frequently in deciduous forests. **Adult**: Brown above and white below streaked and spotted with black on breast. Long rufous tail (vs. grey in feathertoed hawk-eagle). **Immature**: Generally paler, especially on head. Not easily distinguished in the field from juvenile mountain hawk-eagle. (*See* Mountain Hawk-Eagle, 159 for identification in hand)
Call: Rendered as a loud high-pitched cry .. *ki-ki-ki* .. *ki-kee* ...
Range: Resident. Western Ghats up to 1000 m.
Habitat: Deciduous and drier secondary forests and edges of cultivation.

163. Bonelli's Eagle *Hieraaetus fasciatus* (Bonelli's Hawk-Eagle)

Identification: 68–72 cm (female larger). A slender uncrested eagle with long tail extending 5–8 cm beyond the tips of wings at rest. In flight white underparts, blackish underwing contrasting against greyish flight feathers and broad black subterminal band on tail diagnostic. Pairs. **Adult**: Dark brown above (sometimes with a paler mantle). White below streaked with black. Tail dark grey above, whitish below with a conspicuous black sub-

terminal band. Adults in dark colour phase resemble the tawny eagle. **Immature**: Browner and paler. Barring on tail faint. Pattern on underwing unlike the adults and may be confused with buzzards (*Buteo*). However, the longer and narrower neck, wings and tail of hawk-eagles are pointers. (*See* Oriental Honey Buzzard, 130)
Call: Rendered as a chattering .. *kie* .. *kie* .. *kikiki* ...
Range: Resident. Entire Southwestern India.
Habitat: Open forests and occasionally cultivations bordered with tall trees.

164. Booted Eagle *Hieraaetus pennatus* (Booted Hawk-Eagle)

Identification: 50–54 cm (female larger). A small buzzard-like eagle in two distinct colour phases. In overhead flight black wings contrasting with white body reminiscent of the white scavenger vulture. Darker birds resemble marsh harrier and pariah kite. Square-cut longish tail without bands and whitish rump diagnostic. Solitary. **Adult light phase**: Dark umber upperparts with paler shoulders and rump. Buffy-white underparts with blackish streaks. **Adult dark phase**: Dark rufous-brown underparts with black streaks. Pale shoulders and rump. Tail paler. **Immature**: Like adults in dark phase but more rufous.
Call: Rendered as a harsh falcon-like .. *kik-kik-kik* ...
Range: Winter visitor. Entire Southwestern India.
Habitat: Open country, scrub, cultivation and open forests.

165. Rufousbellied Eagle *Hieraaetus kienerii*
(Rufousbellied Hawk-Eagle)

Identification: 53–61 cm (female larger). A short-crested black eagle with white and rufous underparts. Pale patch at base of flight feathers diagnostic in overhead flight. Dives with closed wings like falcon. Solitary. **Adult**: Black above. Chin, throat and upper breast white with black streaks. Rest of underparts rufous-chestnut with black streaks on abdomen and flanks. **Immature**: Brownish-black above. White line across forehead and over eyes. Tail dark brown barred with grey. White underside streaked with black.
Call: Screams rendered as .. *kk-kk-kk-kk-keee-keee* ...
Range: Resident. Western Ghats from Goa to southern Kerala up to 1200 m.
Habitat: Evergreen and moist deciduous forests and their secondary stages.

168, 169. **Eurasian Tawny Eagle** and **Steppe Eagle** *Aquila rapax*
(Tawny Eagle)

Identification: 63–71 cm and 76–80 cm (female larger). A large and heavy-built whitish to brownish-black eagle. Wings long and reaching tip of tail at rest. Tail rounded and broad in flight. Pairs. **Adult:** A highly variable eagle. In flight wings held straight out and level. Underwing lining darker than flight feathers. Race *nipalensis* (steppe eagle) is barely separable in the field from race *vindhiana* (Eurasian tawny eagle). It is, however, larger with a dorsal whitish patch at the base of primary flight feathers. The yellow lining on the gape appears more prominent giving a characteristic facial pattern. Adults are confusable with spotted eagles. **Immature:** Pale plumage and broad white band on underwing unmistakable. Upper tail-coverts with white lining. Lacks conspicuous streaking on the underparts. (*See* Greater Spotted Eagle, 170 and Lesser Spotted Eagle, 171)

NOTE: Some authorities consider the Steppe Eagle as a full species.

Call: Rendered as a guttural .. *kra* .. and grating .. *kekeke* ...
Range: The race *vindhiana* is resident along the drier eastern parts of South-western India as far south as Karnataka (Uttara Kannada) and Tamil Nadu. Race *nipalensis* is a winter visitor recorded from Mumbai.
Habitat: Dry scrub and cultivation.

170. Greater Spotted Eagle *Aquila clanga*

Identification: 64–72 cm (female larger). A large black eagle in the vicinity of water. **Adult:** Blackish-brown. Head occasionally paler. Short rounded tail. In flight the base of wings in some birds with a bulge on rear edge. White lining to upper tail-coverts. Confusable with both tawny eagle and steppe eagle. **Immature:** Shows more white on dorsal surface and above the tail than adults. Finely streaked below.
Call: Rendered as .. *jeb* .. *jeb* .. *jeb* ...
Range: Winter stragglers in Southwestern India recorded as far south as northwestern Karnataka (Uttara Kannada).
Habitat: Large tanks and estuaries.

171. Lesser Spotted Eagle *Aquila pomarina*

Identification: 61–66 cm (female larger). A dark brown-black eagle rather similar to the greater spotted eagle. Smaller build and a preference for drier habitats diagnostic. **Adult:** Blackish. Underwing lining paler than in greater spotted eagle, often paler than flight feathers. Whitish upper tail-coverts less striking. The whitish dorsal patch at base of primaries is rather conspicuous. Separated from adult steppe eagle by less (or lacking) barring on wings and tail and the absence of a yellowish patch on nape. **Immature:**

Medium to dark brown body. Under wing-coverts brown contrasting more or less with grey to blackish-grey flight feathers. Pale patch on under tail-coverts. Yellowish patch on nape. (*See* Eurasian Tawny Eagle and Steppe Eagle, 168, 169)

Call: Recorded as a high-pitched cackling laugh.

Range: Resident in Gujarat and Maharashtra (Mumbai) straggling further south into northwestern Karnataka (Uttara Kannada) and Nilgiris (Kotagiri).

Habitat: Wooded country with scrub and cultivation.

172. Black Eagle *Ictinaetus malayensis*

Identification: 69–81 cm (female larger). A large black forest eagle with tail longer than *Aquila* eagles. In flight wings curved forward near body. Narrowly grey-barred tail, bright yellow cere and feet are further pointers. Soars gracefully. Solitary. **Adult**: Black. Wings reaching tip of tail at rest. Head and nape darker than in tawny eagle and steppe eagle. Pale patch on dark underside of wings conspicuous in flight. At closer range, a white patch under eye diagnostic. **Immature**: Browner underparts. Heavy dark streaks on breast.

Call: Rendered as .. *kip* .. *kip* .. *kip* .. or .. *kee* .. *kee* .. *kee* ...

Range: Resident. Western Ghats up to 2000 m.

Habitat: Evergreen and deciduous forests, open grasslands and patches of montane forests in the higher hills.

173. Whitebellied Fish Eagle *Haliaeetus leucogaster*
(Whitebellied Sea Eagle)

Identification: 66–71 cm (female larger). A large grey and white coastal eagle with black flight feathers. Heavy build and short wedge-shaped tail diagnostic. **Adult**: Grey. Head, underparts and terminal third of tail white. Black primaries and secondaries. **Immature**: Brown with paler (whitish) head. Whitish tail. Black flight feathers. Turns paler and heavily mottled before acquiring adult plumage.

Call: Loud nasal .. *kank* .. *kank* .. *kank* .. *kank* ...

Range: Resident. The entire seaboard of Southwestern India from Mumbai southwards, rarely moving inland up the hills. Vagrants recorded from Gujarat.

Habitat: Beaches, estuaries and tidal rivers. Occasionally inland and upstream within evergreen forests up to 40 km away from the coast.

175. Greyheaded Fish Eagle *Icthyophaga icthyaetus*
(Greyheaded Fishing Eagle)

Identification: 74 cm. A large grey, brown, and white eagle near water. Rounded white tail with black terminal band diagnostic. **Adult**: Grey head

and neck. Brown elsewhere except on belly, flanks and tail which are white. **Immature**: Browner with a white eyebrow. Whitish underparts heavily mottled with brown. In flight pale underside of dark wings and dark band on brown tail diagnostic.
Call: Rendered as loud and clanging reminiscent of the peacock's .. *may-awe* ...
Range: Resident with local movements. Entire Southwestern India.
Habitat: Rivers, lakes, reservoirs and coastal marshes.

178. Redheaded Vulture *Sarcogyps calvus* (King Vulture)

Identification: 84 cm. A medium-sized black vulture with orange-red head, neck, and legs. Tail short and wedge-shaped. Singly or in pairs. **Adult**: Black. In flight wings held in a wide 'V'. White patches on breast and thighs and thin white lining on underside of wings diagnostic. **Immature**: Brown. White lower abdomen and under tail-coverts.
Range: Resident with local movements. Entire Southwestern India along the eastern side up to at least 600 m.
Habitat: Open forest, suburbs, dry cultivation and scrub.

180. Eurasian Griffon *Gyps fulvus* (Griffon Vulture)

Identification: 110–122 cm. An enormous cinnamon-brown vulture with naked head and neck. In flight pale underparts contrasting with black flight feathers and tail diagnostic. **Adult**: Naked head and neck with yellowish down. Pinkish underparts with brown streaks. Whitish ruff at base of neck. **Immature**: Darker brown with brownish ruff. (*See* Longbilled Vulture, 182)
Range: Winter straggler in Gujarat and Mumbai. Once in Dandeli (northern Karnataka).
Habitat: Open semi-desert country. Occasionally suburbs with cultivation and open forest.

182. Longbilled Vulture *Gyps indicus* (Indian Longbilled Vulture)

Identification: 92 cm. A smaller version of the Eurasian griffon vulture with black flight feathers. Dirty white undersides of wings separate from otherwise similar adult whitebacked vulture in flight. Flocks. **Adult**: Pale to dark brown. Head sparsely covered with whitish down. White ruff at base of neck. Underparts pale with dark streaks. **Immature**: Chocolate-brown. Head and neck more thickly clad with down.
Range: Resident with local movements. Southwestern India from Gujarat till at least Jog Falls (Karnataka).
Habitat: Open country and also thin forests bordering tall cliffs.

185. Whiterumped Vulture *Gyps bengalensis* (Indian Whitebacked Vulture)

Identification: 90 cm. A large blackish-brown vulture with bare black head and neck. White back and rump diagnostic. In flight white underwing and contrasting black flight feathers separate from longbilled vulture. Flocks. **Adult:** White ruff at base of neck. White patches on thighs (at rest). **Immature:** Similar to longbilled vulture. Lacks pale back. Separation in mixed flocks difficult except by smaller size.
Range: Resident. Entire Southwestern India.
Habitat: Open and secondary forests, marshes, cultivation and suburbs.

187. Egyptian Vulture *Neophron percnopterus* (White Scavenger Vulture)

Identification: 62 cm. A small white and black vulture with a short wedge-shaped tail. Pairs. **Adult:** Dirty white, yellowish and black plumage. Beak and bare face yellow. **Immature:** Brown with black flight feathers. Longish beak, narrow wings and wedge-shaped tail diagnostic.
Range: Resident with local movements. Entire Southwestern India up to 2000 m in the Nilgiris (Udagamandalam).
Habitat: Dry open country including suburbs and cultivation.

190. Pallid Harrier *Circus macrourus* (Pale Harrier)

Identification: 46-51 cm. A slender, kite-like bird of prey with graceful flight. Grey and white. White upper tail-coverts. Larger than blackwinged kite. Open grasslands and cultivation. **Adult Male:** Pale ashy-grey above. White below. In flight, faintly barred tail, white underparts and black wing-tips diagnostic. **Adult Female:** Umber-brown with an owl-like facial pattern and pale rufous ruff. **Immature:** Similar to female in overall coloration. Unstreaked ruff and underparts. All female and immature harriers are confusingly similar in the field. Best identified in hand. Pallid harrier has tarsus over 65 mm. (*See* Montagu's Harrier, 191)
Range: Winter visitor. Entire Southwestern India up to 2600 m (Nilgiris).
Habitat: Open country, marshes and cultivation, hillside grasslands and scrub.

191. Montagu's Harrier *Circus pygargus*

Identification: 46–49 cm. A brownish dark-grey and white harrier identified by a narrow black wing bar diagnostic at rest and in flight. **Adult Male:** Like pale harrier but darker grey above. Underparts white with narrow rufous streaks. Black wingbar. **Adult Female:** Doubtfully separable in the field from female pale harrier by slightly narrower white upper tail-coverts. **Immature:** Darker than other species with less distinct ruff and facial pattern. In hand, separated from pale harrier by shorter tarsus (less than 65 mm).

Range: Winter visitor. Entire Southwestern India up to at least 1000 m.
Habitat: Swamps, grassy plains, cultivation and open grasslands of the higher hills.

192. Pied Harrier *Circus melanoleucos*

Identification: 46–49 cm. A slim, black and white harrier. **Adult Male**: Black head, mantle, throat and breast. Rump and rest of underparts white. Grey tail. In flight black wing-tips, silvery-white underside and a black dorsal band along secondaries diagnostic. **Adult Female**: Brown and similar to other female harriers. Older females show grey on wings and tail. In flight wings less pointed than in other harriers. Flapping heavier. **Immature**: Dark brown with pale facial pattern. Conspicuous white band on upper tail-coverts. Underparts dark rufous with paler streaks. Confusable with other juvenile harriers. In hand separated from pallid harrier and Montagu's harrier by 2–5 primaries (vs. 2–4) being notched on the outer web. Shorter beak (under 29 mm from cere to tip) distinguishes from marsh harrier.
Range: Winter visitor. Recorded southwards from Mombai as far as the Nilgiris, Palni Hills and sparingly in Kerala up to 2100 m.
Habitat: Grasslands, cultivation, marshes and scrub.

193. Western Marsh Harrier *Circus aeruginosus* (Marsh Harrier)

Identification: 54–59 cm. A largish brown kite-like harrier with pale head and silvery wings and tail. Near water. **Adult Male**: Dark brown with pale (straw-coloured) head, neck and breast. Silvery-grey wings and tail. Black-tipped wings. **Adult Female** and **Immature**: Rather like the pariah kite with unforked tail and pale cap on head. In hand separated from pallid harrier and Montagu's harrier by 2–5 primaries (vs. 2–4) being notched and from pied harrier by longer beak (over 29 mm from cere to tip).
Range: Winter visitor. Entire Southwestern India.
Habitat: Wetlands of all sorts including the salt marshes.

195. Short-toed Snake Eagle *Circaetus gallicus* (Short-toed Eagle)

Identification: 63–68 cm (female larger). A large hawk. Pale colour, broad thickset owl-like head, broad wings and square-cut tail diagnostic. Hovers. **Adult**: Dark brown above. White below streaked and barred with brown. Tail broadly and indistinctly barred, the terminal band being the broadest. Unfeathered legs and upturned bristly feathers on face diagnostic at close range. Thickset head separates from rather similar oriental honey buzzard in flight. **Immature**: Variable in coloration. Head often whitish. White underside with almost unbarred wings and tail.
Call: Plaintive .. *pieeou* .. *pieeou* ...

Range: Resident. Entire Southwestern India up to over 2000 m.

Habitat: Open dry country with scrub and cultivation, coastal plains and the open grasslands of the higher hills.

197. Crested Serpent Eagle *Spilornis cheela*

Identification: 74 cm. A large, black-brown eagle with a full, rounded nuchal crest. In flight white bands on wings and tail diagnostic. Wags tail sideways on alighting. **Adult**: Black and white crest prominent at rest. Yellow base of beak and legs. Underparts with white spots. **Immature**: Paler (whitish) head. Whitish underparts with dark streaks on breast. Confusable with hawk-eagles and short-toed snake eagle. Broad black band through eyes and ear-coverts diagnostic. Tail bands broader and more distinct than those of hawk-eagles. (*See* Oriental Honey Buzzard, 130)

Call: High pitched .. *pi-pi-pieeou-peu-peu* ... Also rendered as .. *kee* .. *kee* .. *kee* .. and *kek-kek-kek-kee*.

Range: Resident. Western Ghats up to 1500 m.

Habitat: Open evergreen, secondary and deciduous forests, scrub and cultivation and occasionally suburbs.

203. Osprey *Pandion haliaetus*

Identification: 56 cm. A large, dark brown and white fishing hawk with a dark band across breast. Angled slender wings. Hovers. Solitary near water. **Adult**: Blackish-brown upperparts with traces of white. White head and neck with dark band through eye and nape. White below with darker breast band and flight feathers. Wing-tips black.

Call: Rendered as .. *kai* .. *kai* .. *kai* ...

Range: Winter visitor. Entire Southwestern India.

Habitat: Rivers, lakes, reservoirs and estuaries.

Family FALCONIDAE

Predatory birds with long pointed and usually narrow wings. Tail narrow and long. Typical facial pattern with dark moustachial streak. Swift direct flight. Females larger than males.

208. Laggar Falcon *Falco jugger*

Identification: 43–46 cm. A large, dark ashy-brown and white falcon with pale head. Moustachial stripe narrow. Unbarred brown tail. In flight dark brown flanks, thighs and underwing diagnostic. Pairs. **Adult**: Ashy-brown upperparts. White below with long pale brown drops; darker on flanks and thighs. Whitish eyebrow. Separated from peregrine falcon by a narrower

moustache, narrower and longer wings (which are also more rounded) and rusty-red nape and hindneck. **Immature**: Overall brown including the underparts. White throat. (See Rednecked Merlin, 219)
Call: Rendered as a shrill prolonged .. *whi-ee-ee* ...
Range: Resident with local movements. Widespread in Southwestern India along the drier side.
Habitat: Dry open country and cultivation, including those along the coast.

209, 211. Peregrine Falcon and Shaheen Falcon *Falco peregrinus*

Identification: 40–48 cm and 38–46 cm. A medium-sized to largish short-tailed streamlined grey-blackish falcon with white or rusty-red and barred underparts. Swift pigeon-like flight diagnostic. **Adult**: Peregrine falcon (race *japonensis*) has bluish-grey upperparts with a distinct black moustachial streak. Underparts whitish with brown barring. Tail indistinctly barred. Shaheen falcon (race *perigrinator*) is smaller. Blackish above and rusty-red below with fine barring. **Immature**: Peregrine falcon is dark brown above and whitish-buffy below with heavy brown streaking except on throat. Shaheen falcon is darker above than adult and more reddish below. (See Laggar Falcon, 208)
Range: Entire Southwestern India. The race *japonensis* is a winter visitor. Race *perigrinator* is resident up to at least 1000 m.
Habitat: Peregrine falcon inhabits the neighbourhood of wetlands including saltmarshes. Shaheen falcon prefers rocky cliffs including those emerging out of forests and those overhanging waterfalls. The latter occasionally visits tall towers within suburbs.

212. Eurasian Hobby *Falco subbuteo* (Hobby)

Identification: 31–34 cm. A small falcon resembling peregrine falcon in colour pattern. Tail shorter. Flight reminiscent of a large swift. **Adult**: Slaty-grey unbarred upperparts. Black moustachial stripe. Rusty-white below heavily streaked with black. Thighs and under tail-coverts unbarred and rufous. Underwing whitish heavily spotted. Dense barring on flight feathers makes them appear grey in the field. The two races are inseparable in the field. **Immature**: Darker (brownish) than adult except on crown. In hand: race *centralasiae* has slightly larger wings averaging not less than 250 mm (vs. 240 mm and above in race *subbuteo*).
Call: Rendered as a harsh plaintive .. *tee-tee-tee-tee-tee* ...
Range: Winter visitor. On the Western Ghats as far south as Belgaum (northern Karnataka).
Habitat: Open forest, cultivation and semi-desert.

214. Oriental Hobby *Falco severus* (Indian Hobby)

Identification: 27–30 cm. A very small falcon—a miniature version of the shaheen falcon. Ferruginous breast and underparts separate from rather similar Eurasian hobby. **Adult:** Blackish head and slaty-grey upperparts including flight feathers. Underparts unstreaked. **Immature:** Like adult but with underparts streaked with black.
Call: Rendered as .. *ki-ki-ki-ki* ...
Range: Winter visitor (?) known from the hills of Kerala.
Habitat: Open forests and woodlands.

219. Rednecked Falcon *Falco chicquera* (Redheaded Merlin)

Identification: 31–36 cm. A medium-sized grey and white falcon with chestnut-red head. Pairs or solitary. **Adult:** Crown, nape, sides of head and cheek-stripe chestnut. Rest of upperparts ashy-grey. Wing-tips blackish. Grey tail finely barred; the terminal band being the thickest with white tips. Whitish below finely barred with black. **Immature:** Duller chestnut on head. More barred (sometimes dorsally as well) than adults.
Call: .. *kip-kip-kip-kip-kip* ...
Range: Resident with local movements. Entire Southwestern India avoiding the coast.
Habitat: Open country and suburbs with tall groves and cultivation.

220. Amur Falcon *Falco amurensis* (Redlegged Falcon)

Identification: 28–31 cm. A small, slender, long-tailed falcon. Sooty with red legs. Batlike flight diagnostic. Hovers. **Adult Male:** Sooty-grey with flight feathers paler above. Thighs and vent red. Legs, cere and skin around eyes orange-red. In flight blackish wings with white lining diagnostic. **Adult Female:** Unstreaked rufous-yellow head. Rest of the upperparts greyish barred with black. Moustachial streak prominent (narrower than in hobby). Lower belly, thighs and vent unbarred. Narrower wings paler on the underside and longer tail separate from similar adult hobby. **Immature:** Like female but browner on back (confusable with common kestrel). Underparts more heavily streaked with black. Dark trailing edge to pale underwing.
Call: Described as shrill screams.
Range: Winter stragglers reported from northwestern Karnataka (Uttara Kannada), Nilgiris and Mumbai.
Habitat: Open country with cultivation and scrub.

221. Lesser Kestrel *Falco naumanni*

Identification: 34 cm. A slim medium-sized falcon with grey head and unspotted rufous back. White underwing. Hovers. Gregarious. **Adult Male:**

Grey head, neck, lower back and rump. Rest of back unspotted rufous. Rufous-buff underparts (unmarked in older birds). Tail ashy-grey with black terminal band and white tip. **Adult Female** and **Immature**: Difficult to separate from female and immature kestrel. However, smaller size, faster flight, taking prey in air, bringing food to beak with foot and at close range whitish claws diagnostic.
Range: Winter visitor. Recorded from Maharashtra and Nilgiris.
Habitat: Open grassy country.

222, 224. Common Kestrel *Falco tinnunculus* (Kestrel)

Identification: 36 cm. A medium-sized slender falcon with pointed wings and long tail. Hovers. Solitary. **Adult Male**: Crown, nape, and sides of neck ashy-grey. Black cheek-stripe. Brick-red mantle with black spots. Tail grey with a black subterminal band. Reddish-buff below, spotted on breast and abdomen with black. Claws black. **Adult Female** and **Immature**: Rufous upperparts streaked and barred with black. Tail rufous with a broad black subterminal band. Moustachial streak not prominent as in male. Underparts heavily spotted or streaked. Older females develop grey on tail (*see* Lesser Kestrel, 221). The races differ in colour. Race *objurgatus* has darker head (often with streaks), mantle richer red than race *tinnunculus* and faintly barred tail.
Call: Rendered as .. *ki-ki-ki-ki* .. or *ti* .. *wee* .. while hovering.
Range: Entire Southwestern India up to 2500 m. The race *tinnunculus* is a winter visitor. Race *objurgatus* is resident above 1200 m.
Habitat: Grasslands on hillsides, dry cultivation and rocky scrub.

Family PHASIANIDAE

The so-called 'game birds'. Ground dwelling. Plumage varies from dull brown to that of a peacock. Males or both sexes often have spurs on legs. Hind toe always present. Three groups are distinguishable. **Quails**: The smallest. Short-tailed, shy and in flocks. **Partridges**: Larger. Longer tail. Mostly in grassy areas. **Pheasants**: Large birds. Brightly coloured including the junglefowl and peafowl. Often with bare face, wattles and ornamental tail plumes.

241. Painted Francolin *Francolinus pictus* (Painted Partridge)

Identification: 31 cm. A medium-sized ground bird. Short tail and squat appearance. Brownish-black with white spots and bars. Spurs absent in both sexes. In flight black outer tail feathers and rufous on wings diagnostic.
Adult Male: Supercilium and face pale chestnut. Black below spotted with

white. Centre of belly rufous. **Adult Female**: often has whitish throat.
Immature: Flanks and belly buffy with black arrow-shaped marks.
Call: High pitched .. *click* .. *cheek* .. *cheek* .. *keray* .. repeated in quick suc-
cession.
Range: Resident. Raigad district in Maharashtra and Uttara Kannada–Shimoga
in Karnataka.
Habitat: Open grasslands and scrub in secondary forests.

246. Grey Francolin *Francolinus pondicerianus* (Grey Partridge)

Identification: 33 cm. A medium-sized village hen-like ground bird. Greyish-
brown with darker and white streaks and mottling. Rufous tail. Pairs or
small flocks along roadsides and cultivation. **Adult**: Brown, rufous and
chestnut above mottled with black and white. Buff below finely cross-
barred with black. Yellowish throat patch bordered by black. In flight
rufous tail prominent. **Male**: With spurs on legs. **Immature**: Duller.
Call: High pitched .. *kateetar* .. *kateetar* .. *kateetar* .. (male) and .. *kila* .. *kila*
.. *kila* .. (female). Mated pairs sing a duet rendered as .. *kateela* .. *kateela*
kateela ...
Range: Resident. Entire Southwestern India along the foothills and up to 1000 m.
Beyond Pune it is likely that the southern race *pondicerianus* mingles with
or is displaced by the northern race *interpositus*.
Habitat: Open grassland, cultivation and scrub.

250. Common Quail *Coturnix coturnix* (Grey Quail)

Identification: 20 cm. A small ground bird with bold buffy streaks on upperparts and prominent whitish eyebrow. Pairs. **Adult Male**: Chin and a
line down the centre of throat black. Breast appears unmarked and buffy.
Flanks brown and streaked with black. **Adult Female** and **Immature**: Chin
and throat creamy-white. Rest of underparts including breast heavily
spotted with black. (*See* Rain Quail, 252)
Call: Squeaky whistle when flushed. Call rendered as .. *wet-mi-lips* .. uttered
in the mornings, evenings and sometimes during nights.
Range: Resident as far south as Satara (Maharashtra). Winter visitor over the
rest of Southwestern India.
Habitat: Grasslands and cultivation.

252. Rain Quail *Coturnix coromandelica* (Blackbreasted Quail)

Identification: 18 cm. A small quail rather similar to common quail. Black
patch on breast. Black lines on face more prominent. Singly or scattered
pairs. **Adult Male**: White throat with a broad black line down centre and
two black bands. Black breast. Sides and flanks cinnamon streaked with
black. **Adult Female**: Similar to female common quail except smaller size.

In hand the unbarred primary feathers (vs. barred in common quail) are diagnostic for both sexes.

Call: Distinct and rendered as .. *which-which* .. *which-which* .. uttered repeatedly in the mornings, evenings and during overcast days.

Range: Resident with local movements. Entire Southwestern India.

Habitat: Grasslands, scrub and cultivation.

253. Bluebreasted Quail *Excalfactoria (Coturnix) chinensis*

Identification: 14 cm. A very small, dark quail with yellow legs. Female plumage very different. Pairs or small flocks in wet areas. **Adult Male**: Forehead, supercilium and sides of neck slaty-blue. Rest of upperparts and throat like grey quail. Breast and flanks slaty-blue. Abdomen and under tail-coverts rich chestnut. **Adult Female** and **Immature**: Like female common quail. Broader pale eyebrow. Black barring on underparts (vs. longitudinal streaks in common quail) diagnostic.

Call: Rendered as a soft .. *tir* .. *tir* .. *tir* .. when flushed. Other calls have been described as resembling peeping whistles and those of the greybellied cuckoo and common hawk-cuckoo.

Range: Resident with local movements. Widespread south of Mumbai up to 2000 m in the Western Ghats.

Habitat: Swampy grasslands and scrub.

255, 257, 257a. Jungle Bush Quail *Perdicula asiatica*

Identification: 16 cm. A small, partridge-like quail with short curved beak. Bright chestnut throat diagnostic. Flocks. **Adult Male**: Brown above streaked and mottled with black. White below closely barred with black. Prominent buff and chestnut superciliary stripe from forehead and sides of neck. **Adult Female**: similar to male except for the unbarred underparts. **Immature**: Duller. Streaked with creamy-white both above and below. (*See* Rock Bush Quail, 260). The races vary in intensity of brown on plumage. Race *asiatica* has whitish shaft streaks on upperparts. Race *vidali* is more reddish on back. Race *vellorei* is paler on back with a dark chocolate-brown chin and throat.

Call: Flocking calls rendered as .. *whi-whi-whi-whi-whi-whi* .. and challenging crows of the males as .. *chee-chee-chuck* .. *chee-chee-chuck* .. reminiscent of the black drongo.

Range: Resident. Entire Southwestern India up to 1500 m. The race *vidali* occurs in the Western Ghats. Mixes with races *asiatica* and *vellorei* in the east.

Habitat: Grasslands, scrub, secondary and deciduous forests.

260. Rock Bush Quail *Perdicula argoondah*

Identification: 17 cm. A small thickset quail similar to jungle bush quail. Male has no pale supercilium. Female lacks the red throat patch. **Adult Male**: Brick-red throat patch (vs. rufous in jungle bush quail) diagnostic. Black streaks on underparts. **Female**: Chin whitish. Throat, breast and underparts unstreaked pinkish-buff. **Immature**: Differs from immature jungle bush quail in having streaked underparts as in the adult males.

Call: Described as similar to that of jungle bush quail.

Range: Resident. Entire Southwestern India. The race *argoondah* is found over most of the peninsula though apparently replaced by race *meinertzhageni* in Saurashtra (Gujarat) and race *salimalii* in Wynaad district, Kerala.

Habitat: Stony semi-desert country with scrub.

262. Painted Bush Quail *Perdicula erythrorhyncha*

Identification: 18 cm. A small brightly coloured quail of higher elevations. Deep red beak and legs diagnostic. **Adult Male**: Olive-brown above finely streaked with white and boldly blotched with black. White stripe running down from over eye. Throat white bordered with black. Chestnut underparts mottled boldly with black and white on flanks. **Adult Female**: Like male. Pattern on head indistinct. Flanks less strikingly blotched with black.

Call: Rendered as .. *tu-tu-tu-tu-tutu-tutu-tuttu* .. and also as .. *kirikee* .. *kirikee*...

Range: Resident. Khandala (Maharashtra) through Coorg, Nilgiris and Kerala between 600 and 2000 m.

Habitat: Tall grass and grassland interspersed with low scrub and cultivation.

275, 277. Red Spurfowl *Galloperdix spadicea*

Identification: 36 cm. A reddish-brown village hen-like ground bird with longish flattened tail. Red face and legs with many spurs diagnostic. Pairs or small flocks. **Adult Male**: Brown-blackish crown. Dark brown and rufous-chestnut upperparts with black vermiculations. Chin whitish. Rest of the underparts rufous-chestnut scalloped grey-brown. Bare patch on face brick-red. Two to four pointed spurs on each leg. **Adult Female**: Sandy or greyer than male. Underparts spotted with black. Naked face paler. One or two spurs on each leg. **Immature Male**: Similar to female. Richer coloration with more black in plumes. The races differ chiefly in colour. Male race *spadicea* is duller chestnut with brown crown; race *stewarti* being brighter chestnut with black crown. Female, of race *stewarti* is generally darker and more rufescent.

Call: Cackling call rendered as .. *kuk-kuk-kuk-kukaak* .. Cocks crow .. *kr-r-r-kwek* .. *kr-kr-kwek* .. *kr-kr-kwek* ...

Range: Resident. Entire Western Ghats up to 1250 m. The northern race *spadicea* is separated from race *stewarti* at the Palghat gap in Kerala.
Habitat: Open evergreen and deciduous forests, secondary scrub with *Lantana, Eupatorium* and bamboo and occasionally in the outskirts of hill towns and estates.

278. Painted Spurfowl *Galloperdix lunulata*

Identification: 32 cm. A metallic green-black and chestnut spurfowl distinguished from red spurfowl by its brighter coloration, white spots on plumage and absence of a bare red patch around eyes. Underparts buff and chestnut stippled with black triangles. Pairs or flocks in dry scrub. Adult Male: Black head, neck, wings and tail. Adult Female: Duller. Black only on crown. Brown wings and tail. Underparts paler and indistinctly mottled with chestnut. Buff moustachial streak diagnostic. Immature Male: Like adult female but entire plumage much more freckled and mottled.
Call: Described as domestic fowl-like.
Range: Resident. Locally on the eastern foothills of the Western Ghats. Recorded from southern Tamil Nadu.
Habitat: Grass and thorny scrub with bamboo.

299. Red Junglefowl *Gallus gallus*

Identification: 43–66 cm (with long tail). A medium-sized, red and metallic black 'village cock' with red wattles and pale lappets. Pairs or flocks. Adult Male: Glossy orange-red above with yellowish hackles and elongate feathers on rump. Black tail with long sickle-shaped feathers. Blackish underparts. Adult Female: Paler rufous crown and head. Rest of plumage dark brown with buff and black streaks. Naked face red. (*See* Grey Junglefowl, 301)
Call: Crow described as similar to a small village cock.
Range: Resident. As far south and west as Khandala (Maharashtra) in the Western Ghats.
Habitat: Deciduous forest and secondary scrub with cultivation.

301. Grey Junglefowl *Gallus sonneratii*

Identification: 46–80 cm (with long tail). A medium-sized fowl with long tail. Grey, yellow and black plumage diagnostic. Pairs or small flocks. Male breeding: Grey with glossy black wings and tail including the long sickle-feathers. Hackles, shoulders and upper back golden-yellow. Blackish below streaked with white. Bare face, wattles and lappets red. Female: Grey-brown finely streaked with black and white above. Whitish below (vs. brown in red junglefowl)), scalloped with black on breast. Immature

and **non-breeding Male**: Duller than breeding male with less yellow on hackles and shorter tail. Fleshy comb reduced.
Call: .. *Kuck-kaya-kaya-kuck* .. *kuk-ka-kaya-kuk* .. and .. *klick* .. *kluck-kluck* ...
Range: Resident. Western Ghats up to over 2000 m, overlapping with *G. gallus* along the northern limits of its range.
Habitat: Evergreen and deciduous forests, secondary open forests and scrub, all kinds of hill plantations, including eucalyptus, grasslands, hillside cultivation, bamboo facies and outskirts of villages.

311. Indian Peafowl *Pavo cristatus*

Identification: 86–122 cm (males 250 cm with full 'train'). A large, familiar blue, green and brown pheasant with a fan-like crest. Solitary or in flocks. **Male breeding**: Typical peacock with magnificent upper tail-coverts (train). **Female**: Smaller and browner. Greenish neck. Whitish below. **Immature** and **non-breeding Males**: Larger and brighter than female with more blue on neck and chestnut on wings. Lack the long upper tail-coverts.
Call: .. *may-awe* ... Goose-like .. *ka-an* .. *ka-an* ...
Range: Resident. Entire Southwestern India up to 1800 m.
Habitat: Deciduous forests, secondary forests with grass and scrub, cultivation and around villages.

Family TURNICIDAE

Small ground birds resembling true quails. Feet with only three toes. Females larger than males and more brightly coloured. Booming calls.

NOTE: Sibley and Monroe (1990) treat Button Quails as belonging to a separate order, viz. TURNICIFORMES.

313. Small Button Quail *Turnix sylvatica* (Little Bustard Quail)

Identification: 13 cm. A very small quail with short pointed tail. Legs and feet not yellow. **Adult**: Blackish crown with a paler central line. Upperparts (especially upper wing-coverts) streaked. Underparts whitish. Breast buff with black and chestnut spots on its sides. (*See* Yellowlegged Button Quail, 314)
Range: Resident with local movements. South up to Kerala.
Habitat: Grass and scrub jungle.

314. Yellowlegged Button Quail *Turnix tanki*

Identification: 15 cm. A small quail with yellow legs. Barred wings and shoulders diagnostic. Breast darker rufous-buff than small button quail.
Range: Resident with local movements. Entire Western Ghats up to 1200 m.
Habitat: Damp grasslands, scrub, bamboo facies and cultivation.

318. Barred Button Quail *Turnix suscitator* (Common Bustard Quail)

Identification: 15 cm. A small quail with grey beak and feet. White eyes. Barred breast. Pale shoulders diagnostic in flight. **Male**: Confusable with female bluebreasted quail. Bright buffy breast, broader black bands and unmarked rich rufous-buff lower flanks, thighs and under tail-coverts diagnostic. **Female**: Black chin, throat and upper breast diagnostic. Sides of breast and flanks barred with black. (*See* Rain Quail, 252)
Range: Resident with local movements. Entire Southwestern India up to over 2000 m.
Habitat: Grassland, scrub jungle and open forest.

Family GRUIDAE

Large marsh birds with long legs and neck. Crown and nape often bare. Long feathers drooping over tail at rest. Neck outstretched in flight. Loud, trumpeting calls. Sexes alike.

323. Sarus Crane *Grus antigone*

Identification: 156 cm. A very large, long-necked marsh bird. Grey plumage with red head diagnostic. **Adult**: Crown, and nape bare. Black flight feathers. Female smaller. **Immature**: Head covered with buffy-brown feathers.
Range: Resident with local movements. Gujarat.
Habitat: Marshes and cultivation.

326. Demoiselle Crane *Grus virgo*

Identification: 76 cm. A small grey crane with black head and neck. White ear-tufts diagnostic. Flocks **Adult**: Long black feathers of neck fall over breast. Sickle- shaped wing feathers fall over tail. Black flight feathers. **Immature**: Greyish head and neck. Shorter ornamental plumes.
Range: Winter visitor. South up to Karnataka along the eastern side.
Habitat: Cultivation, edges of marshes and reservoirs.

Family RALLIDAE

Marsh birds capable of wading as well as swimming. Many are crepuscular. Walk jerking their heads and flicking tails. Flight is weak with legs dangling behind. **Rails** and **Crakes** are shy and secretive. **Waterhen, Watercock, Moorhen** and **Coot** are easily seen out of cover. In flight appear small-headed and long-necked.

329. Slatybreasted Rail *Gallirallus striatus* (Bluebreasted Banded Rail)

Identification: 27 cm. A medium-sized short-tailed marsh bird with longish beak skulking among reeds. Brown and grey with white and black bands. Sexes alike. **Adult**: Grey breast. White chin and throat. White bands and spots above and black and white bands on the underparts diagnostic. Reddish beak and eyes. Greyish legs. **Immature**: Dark streaks above; the white dots and bars are absent or faint. Less distinct barring on underparts.
Range: Resident with local movements. Entire Southwestern India.
Habitat: Reedy swamps, paddyfields and mangroves.

332. Slatylegged Crake *Rallina eurizonoides* (Banded Crake)

Identification: 25 cm. A medium-sized cinnamon-brown rail with no white bars on back. Slaty-grey legs. Sexes alike. **Adult**: White bars on wings visible in flight. Black and white barring on underparts. **Immature**: Olive-brown. Shoulders with sparse white dots and bars. Legs slaty.
Call: Rendered as .. *kek-kek* .. *kek-kek* .. *kek-kek* ... Heard during nights.
Range: Resident with local movements. Western Ghats up to 1600 m.
Habitat: Dense, well-watered jungles.

336. Little Crake *Porzana parva*

Identification: 20 cm. A small, rufous and grey marsh bird, finely streaked with black and white above. Bold white bars below. Green legs. **Adult Male**: Olive-brown above with grey and rufous on head. White streaking above is feeble. Grey below with the white bands restricted to the tail end. **Adult Female** and **Immature**: White chin, throat and foreneck. Buff (almost whitish) underparts barred with brown and white on vent and under tail-coverts. (*See* Baillon's Crake, 337)
Range: Winter visitor. Known from Mumbai and coastal Uttara Kannada (Karnataka).
Habitat: Reed-beds.

337. Baillon's Crake *Porzana pusilla*

Identification: 19 cm. A very small marsh bird with short greenish beak. Olive-brown with fine black and white streaks above. Barred underparts including flanks. **Adult**: Brown line through eyes and ear-coverts. White edge to shoulder conspicuous in flight. Grey underparts. **Immature**: Brown eye-streak broader. Whitish underside with faint brown barring. (*See* Spotted Crake, 338)
Range: Winter visitor. Entire Southwestern India.
Habitat: Reed-beds and irrigated fields.

338. Spotted Crake *Porzana porzana*

Identification: 23 cm. A small grey and brown swamp bird with white specks on head and neck. White underparts with specks on breast and flanks diagnostic. **Adult**: Head and neck largely grey with white spots. Rest of upperparts olive-brown with black streaks and towards the rump streaked white. Chin, throat and abdomen white. Breast greyish speckled with white. Flanks brownish barred white. Dirty white under tail-coverts. White edge to folded wings. **Immature**: More white on chin and throat. Rest of underparts browner. (*See* Baillon's Crake, 337 and Little Crake, 336) **Range**: Winter visitor. Widespread in Southwestern India as far south as Belgaum in northern Karnataka. **Habitat**: Reedy marshes.

340. Ruddybreasted Crake *Porzana fusca* (Ruddy Crake)

Identification: 22 cm. A small chestnut and olive unbarred rail with white chin and throat. Dark under tail-coverts. Red legs dangling diagnostic in flight. Sexes alike. **Adult**: Dark under tail-coverts with white fringes. White barring on wings and belly feeble or absent. **Immature**: Darker than adult. White on face. Underparts barred and whitish except on the under tail-coverts which are dark as in adults. **Range**: Resident with local movements. Mumbai to Kerala up to 2000 m in Southwestern India. **Habitat**: Reed-covered marshes and paddyfields.

344. Whitebreasted Waterhen *Amaurornis phoenicurus*

Identification: 32 cm. A medium-sized blackish and white marsh bird with reddish under tail. Tail-flicking diagnostic. Sexes alike. **Adult**: Dark slaty-grey. White face, breast and abdomen. Rufous under tail-coverts. **Immature**: Duller lacking the sharp contrast between the dark upperparts and white underparts. **Call**: Croakings like frogs in a chorus rendered as .. *krr-kwaak-kwaak* .. *krr-kwaak-kwaak* .. or .. *kook* .. *kook* .. *kook* ... Often heard on overcast and rainy days incessantly for 10–15 minutes. **Range**: Resident. Entire Southwestern India up to 2000 m. **Habitat**: Wetlands of all kinds, including paddyfields and urban sewage drains and drier cultivation.

346. Watercock *Gallicrex cinerea*

Identification: 43 cm. A largish dark grey-brownish marsh bird scalloped paler, creating a scaly effect. A small yellowish horny shield on forehead diagnostic. Female smaller. **Male breeding**: Black scalloped with grey.

Bright red fleshy horn on forehead, eyes and legs. **Non-breeding Male,
Female** and **Immature**: Grey with a scaly pattern. Reduced yellowish
fleshy horn on forehead. Greenish legs.
Call: Booming calls rendered as .. *kok-kok-kok-kok* .. *kok* .. (10–12 times) followed by .. *utumb-utumb-utumb* .. and .. *kluck-kluck-kluck-kluck-kluck* ...
Range: Resident with local movements. Entire Southwestern India.
Habitat: Wetlands, especially those with emergent aquatic plants.

347. Common Moorhen *Gallinula chloropus*

Identification: 32 cm. A medium-sized black rail with red beak appearing
like a duck in water. Sexes alike. **Adult**: Black above. White border to
closed wings. Slaty-grey below. White under tail-coverts parted with black.
Red frontal shield and beak; the latter with yellow tip. Greenish legs. **Immature**: Brownish with more white on the underparts. Beak and frontal
shield greenish.
Call: Chuckling notes rendered as .. *kirrik-krek-rek-rek* ... uttered constantly.
Range: Resident. Entire Southwestern India up to 2000 m (Nilgiris).
Habitat: Marshes and tanks with emergent aquatic vegetation.

349. Purple Swamphen *Porphyrio porphyrio* (Purple Moorhen)

Identification: 43 cm. A large, blue and purple marsh bird with white under
tail-coverts. Red beak, forehead and legs. Blackish in poor light. Tail-flicking and large size diagnostic. **Adult**: Unmistakable for its size and blue
coloration. **Immature**: Duller with blackish beak and forehead. Orange-brown legs and feet.
Call: Noisy cackling and hooting.
Range: Resident with local movements. Entire Southwestern India.
Habitat: Reed-beds, paddyfields and larger tanks with emergent aquatic
vegetation.

350. Common Coot *Fulica atra*

Identification: 42 cm. A largish black duck-like marsh bird with white beak
and forehead, swimming and skittering in open water. Gregarious and
diving. **Adult**: Black. White trailing edge to wing diagnostic in flight. Legs
green. **Immature**: Dull slaty-grey-brown. Whitish below. No white on
under tail-coverts. Beak and forehead not white.
Call: A nasal .. *henk* ... Also heard at night.
Range: Resident with local movements. Entire Southwestern India.
Habitat: Lakes, reservoirs, reedy tanks and estuaries.

Family OTIDIDAE

Large ground birds with long legs. Three-toed. Long necks with occasional ornamental plumes from behind eyes. Body horizontal to ground while walking. Flight powerful on broad rounded wings.

357. Lesser Florican *Eupodotis indica*

Identification: 46–51 cm. A largish long-legged and long-necked ground bird. Sandy-brown mottled with black arrowhead-like marks on back. Female larger. **Male breeding**: Black and white with a tuft of upturned long plumes from behind eyes. **Non-breeding Male, Female** and **Immature**: Mottled brown. White on wings (male).
Range: Resident with local movements. Widespread in Southwestern India as far south as Karnataka. Stray records from Nilgiris (1000 m) and Kerala.
Habitat: Tall grassland, scrub and cultivation.

Family JACANIDAE

Long-legged marsh birds with extremely long toes. Walk on floating plants. Brightly coloured and sometimes with ornamental breeding plumes. Capable of swimming. Legs dangle weakly in flight. Sexes alike.

358. Pheasant-tailed Jacana *Hydrophasianus chirurgus*

Identification: 31 cm (+ 25 cm tail at breeding). A long-legged yellowish marsh bird. Black-tipped white wings diagnostic in flight. On floating plants. **Breeding**: Head and neck white. A black bordered yellow patch on nape. Brown upperparts. Blackish-brown underparts and long pheasant-like tail. **Non-breeding**: Pale brown and white with black necklace on breast. Long tail absent. **Immature**: Like non-breeding adult. More yellowish. Lacks the black necklace.
Call: Shrill calls rendered as .. *tewn-tewn* .. (as the birds take off) and .. *me-e-ou* .. *me-e-ou* ...
Range: Resident with local movements. Entire Southwestern India.
Habitat: Tanks and small lakes with floating vegetation.

359. Bronzewinged Jacana *Metopidius indicus*

Identification: 28 cm. A glossy black and bronze long-legged marsh bird on floating vegetation. Stubby tail. White eyebrow. In flight, black wings and dangling legs diagnostic. **Adult**: Bronzy-olive wings. Rest of plumage glossy black except on eyebrow. Reddish rump and tail. **Immature**: Brownish

head and whitish underparts. Dark wings separate from immature pheasant-tailed jacana in flight.
Call: Shrill squeaks rendered as .. *seek-seek-seek* ...
Range: Resident. Entire Southwestern India.
Habitat: Lakes and ponds with floating vegetation and flooded rice fields.

Family HAEMATOPODIDAE

Large, coastal waders. Black and white plumage. Red legs and slightly upturned beak distinctive. Sexes alike.

360. Eurasian Oystercatcher *Haematopus ostralegus* (Oystercatcher)

Identification: 42 cm. A largish black and white shorebird with red legs and beak. In flight white band on wings, white rump and underparts contrasting with black body diagnostic. **Non-breeding Adult** and **Immature**: White patch on throat. Black on plumage browner in immature birds.
Call: High pitched, two-note whistle. In flight .. *kleep* ...
Range: Winter visitor along the entire seaboard.
Habitat: Sandy and rocky beaches.

Family CHARADRIIDAE

Long-legged, short-beaked shorebirds. Thick-headed. Prefer open ground near water. Run fast. **Lapwings** are larger with broader wings and white tails. **Plovers** are smaller with narrower wings, faster flight. Sexes alike.

362. Whitetailed Lapwing *Vanellus leucurus*

Identification: 28 cm. A medium-sized pale brown long-legged bird with white face and grey breast. In flight white and black wings and pure white tail diagnostic. Near water. **Adult**: Black flight feathers. Lower back and upper tail-coverts white. Yellow legs. **Immature**: Brownish boldly mottled with black. White tail with brownish subterminal band. (*See* Sociable Lapwing, 363)
Call: Rendered as .. *pee-wick* .. or .. *kwie-wuck* ...
Range: Winter visitor into Gujarat straggling as far south as Mumbai.
Habitat: Edges of tanks and lakes.

363. Sociable Lapwing *Vanellus gregarius*

Identification: 33 cm. A medium-sized plain brown long-legged bird with dark crown and whitish eyebrow. Black and white wings and white tail with broad black tips diagnostic in flight. Black legs. Gregarious on dry ground. **Breeding**: White forehead and long supercilia. Black stripe through eyes. Yellowish chin and throat. Grey breast and black belly. White under tail. **Immature**: Brown like non-breeding adult. White supercilium and scaly (not mottled as in immature whitetailed lapwing) pattern on back diagnostic.

Range: Winter visitor to Gujarat and Maharashtra, straggling down to Kerala.

Habitat: Dry wasteland, ploughed and harvested fields.

365. Greyheaded Lapwing *Vanellus cinereus*

Identification: 37 cm. A largish, heavy-built long-legged bird with grey head, neck and breast. Brown mantle and white underparts. Black and yellow beak. Yellow legs. Black and white wings and black-tipped white tail diagnostic in flight. Solitary or gregarious. **Non-breeding**: Brown head and neck, tinged grey. Chin and throat whitish streaked brown. Greyish breast bordered with an ill-defined blackish band. **Immature**: Similar to non-breeding adult. Scaly pattern on back. Bicoloured beak diagnostic. (*See* Whitetailed Lapwing, 362 and Sociable Lapwing, 363)

Call: Rendered as a plaintive .. *chee-it* ...

Range: Winter visitor straggling south. One sight record from northeastern Uttara Kannada (Karnataka).

Habitat: Edge of marshes and on tank beds.

366. Redwattled Lapwing *Vanellus indicus*

Identification: 33 cm. A largish brown, white and black long-legged bird with red wattle and beak. Yellow legs. Characteristic calls diagnostic. Near water. **Adult**: Black head and neck. White band from below the eyes running down through the breast. Bronzy-brown back. White below. Black-tipped beak. White wings and tail with broad black tips visible in flight. **Immature**: Brownish head. Chin, throat and neck largely white. Red beak diagnostic. (*See* Yellow-Wattled Lapwing, 370)

Call: Familiar .. *did-he-do-it* .. or .. *did-did-did-did* .. at rest and in flight.

Range: Resident. Entire Southwestern India up to 2000 m (Nilgiris).

Habitat: Wetlands of all kinds, river and stream banks, salt-marshes and cultivation.

370. Yellow-wattled Lapwing *Vanellus malabaricus*

Identification: 27 cm. A medium-sized brown and white long-legged bird. Black cap, yellow wattle and legs diagnostic. Pairs or small flocks on dry ground. **Adult**: Black cap separated from brown body by white line. White belly. Black-tipped white wings and tail. Smaller size, black cap, black beak and narrow white bar on dorsal surface of wing (in flight) separate from similar greyheaded lapwing. **Immature**: Sandy-brown above mottled finely. Whitish below. Black beak. Black tips on white tail. (*See* Greyheaded Lapwing, 365)

Call: High pitched .. *ti-ee* .. *ti-ee* .. and .. *tit-tit-tit-tit* .. in flight. Heard at nights also.

Range: Resident with local movements. Entire Southwestern India.

Habitat: Dry cultivation, grassland, scrub and riverbeds.

371. Grey Plover *Pluvialis squatarola*

Identification: 31 cm. A medium-sized mottled grey wading bird. Hunched stance. Black armpits, white rump and wingbar (above) diagnostic in flight. Flocks. **Non-breeding**: Brownish-grey above finely mottled with white. Dark line through eye and behind ears. White below mottled with brown. Thick black beak. Black legs. **Breeding**: Black underparts except vent and tail which are white. A white band from forehead along sides separating grey back from black underparts. (*See* Eastern Golden Plover, 373)

Call: Loud .. *tlee-oo-ee* ... Often heard when the birds take off.

Range: Winter visitor. Entire Southwestern India.

Habitat: Sandy shores, tidal mudflats, salt marshes and harvested rice fields.

373. Pacific Golden Plover *Pluvialis fulva* (Eastern Golden Plover)

Identification: 24 cm. A mottled wader, smaller and slimmer than the grey plover with a golden-yellow touch on the entire plumage. Smoky-grey underwing, absence of white wingbar and fanned-out tail diagnostic in flight. **Non-breeding**: Brown above mottled and fringed with bright yellow. Eyebrow yellowish. Brownish underparts washed with bright yellow. Black beak and legs. **Breeding**: Blackish-brown above spangled with golden and white. Black below. A white line from forehead along sides separating the black underparts from the golden-brown back.

Call: Sharp calls rendered as .. *tu-ee* .. or .. *chu-wit*... Single whistle .. *teeh* .. or *kl-ee* ...

Range: Winter visitor. Entire Southwestern India along the plains and seaboard.

Habitat: Sandy shores, tidal mudflats, salt-marshes, harvested rice fields and edges of larger tanks or lakes.

374. Greater Sandplover *Charadrius leschenaultii* (Large Sandplover)

Identification: 22 cm. A long-legged, short-billed, hunched, sandy-brown and white shore bird lacking a white collar. Larger size, thickset pigeon-like head, longer beak and paler legs separate from the very similar Mongolian sandplover at rest. In flight more white on wings, sides of rump and tail and a more contrasting dark subterminal band on tail are diagnostic. **Adult non-breeding**: Sandy-brown with a scaly pattern above. White forehead,

eyebrow, chin and throat. Brown band across breast often joining narrowly at the centre. White below. Black beak and greenish-grey legs. White wingbar visible in flight. **Adult breeding**: Greyish crown and nape with broad black band from forehead through eyes. White patch on forehead. White chin and throat. Chestnut sides of neck and foreneck forming a band across the breast. Sandy-brown back. White underparts. In all plumages easily confused with Mongolian sandplover that occurs alongside.
Call: Soft note rendered as ,. *trrri* ...
Range: Winter visitor along the entire seaboard.
Habitat: Sandy shores, tidal mudflats and salt marshes.

376. Caspian Sandplover *Charadrius asiaticus*

Identification: 19 cm. Sandy-brown and white shore bird with long greenish legs and short, slender, black beak. Prominent whitish supercilium and a broad brown band across breast. Short white wingbar on inner primaries, whitish underwing with darker armpits and short squarish whitetipped tail diagnostic in flight. **Adult non-breeding**: Sandy-brown above with a scaly pattern. Forehead, supercilium, chin and throat white. Brown band across breast (mottled in immature birds). Rest of underparts white. **Adult breeding**: Almost white face. Breast band rufous bordered below with black.
Call: Shrill whistles rendered as .. *ku-wit* .. or .. *tyup* ...
Range: Vagrant. Maharashtra coasts.
Habitat: Seaboard.

379, 380. Little Ringed Plover *Charadrius dubius*

Identification: 17 cm. Small brownish and white plover running about with typically hunched back. A single breast band. Black beak and yellow legs. Yellow ring around eye. No wingbar visible in flight. **Adult breeding**: White forehead. Forecrown black separated from brown hindcrown by a white line. Black band through yellowrimmed eyes. Black and white collar on hindneck; the black running across the breast forming a band. Rest of plumage brown above and white below. **Adult non-breeding** and **Immature**: Overall sandy-brown and white; the face and breast pattern browner. Lack of white bar on wing separates from similar kentish plover in comparable plumage. The two races are not easily separated in the field except by the colour of the beak and length of wings. Race *curonicus* has longer wings. Race *jerdoni* has more yellow on the base of beak.
Call: Shrill .. *phieou* .. or .. *pee-oo* .. in flight.
Range: Race *curonicus* is a winter visitor and race *jerdoni* is resident with local movements. Entire Southwestern India along the plains and seaboard.
Habitat: River banks, beaches, tidal mudflats and damp fields bordering estuaries.

381. Kentish Plover *Charadrius alexandrinus*

Identification: 17 cm. Small, long-legged plover with the characteristic hunchback. Pale sandy-brown and white with a distinct white collar across hindneck in all plumages. Incomplete breast band. Black legs. In flight, white wingbar and outer tail feathers diagnostic. **Adult breeding**: White forehead and supercilium. Forecrown and band through eyes black. Rufous-chestnut crown. Black legs. Rest of plumage sandy-brown above and white below. **Non-breeding** and **Immature** birds are separated from Mongolian sandplover in comparable plumage by the distinct white collar.
Call: Flight call rendered as .. *pit* .. or .. *twit* ... Others are .. *prrr* .. and .. *too-eet* ...
Range: Resident in Gujarat. Winter visitor elsewhere. Entire seaboard. Occasionally inland.
Habitat: Beaches, tidal mudflats, salt marshes and salt pans.

384. Mongolian Sandplover *Charadrius mongolus* (Lesser Sandplover)

Identification: 19 cm. A plain brown and white miniature of greater sandplover. Dark grey and shorter legs with toes that do not project beyond tail and the white wingbar of even width are pointers to separating lesser sandplover from large sandplover in flight. **Adult breeding**: Differs from greater sandplover in having a broader chestnut band on breast which also spreads over the sides. **Immature**: Similar to non-breeding adults in general. Upperparts show a scaly pattern. Lack of white collar on hindneck separates from kentish plover in comparable plumage.
Call: Rendered as .. *trr* .. or .. *twip* .. or .. *chitik* .. while taking off.
Range: Winter visitor all along the seaboard.
Habitat: Beaches, tidal mudflats and fields along estuaries.

Family SCOLOPACIDAE

Long-legged shore birds with slender and usually long beaks. Mostly seen in winter in confusingly similar mottled grey / brown and white plumages. Gregarious. Noisy and fast-flying. **Curlews** and **Whimbrels**: large with long downcurved beaks. **Godwits**: large with long slightly upturned beaks. **Turnstones**: stout, short beaks, short legs and bright colour patterns. **Snipes** and **Woodcocks**: long beaks, cryptic coloration, secretive and erratic flight. **Sandpipers** (*Tringa, Actitis, Calidris, Limicola* and *Philomachus*): short-long beaks, straight or decurved, greyish mottled and white plumage, curious bobbing while foraging and solitary or gregarious.

385. Whimbrel *Numenius phaeopus*

Identification: 43 cm. A fairly large long-legged grey-brown shore bird with long downcurved beak. Pale supercilium diagnostic. Gregarious. **Adult**: Mottled grey-brown with darker crown. Characteristic crown pattern of paler mid-stripe and supercilia diagnostic. Underparts white with brown-grey streaks and spots. White lower back and rump and mottled brownish tail visible in flight. **Immature**: Like adults in general. Wings with buff spots and breast more buff tinged.
Call: A continuous musical .. *bibibibibi* .. or .. *tetti-tetti-tetti-tetti* .. *tetti* ...
Range: Winter visitor. Entire seaboard.
Habitat: Beaches, mangrove and estuaries with emergent reeds.

388. Eurasian Curlew *Numenius arquata*

Identification: 58 cm. Large. Sandy-brown, mottled black and white. Long down-curved beak. Lack of any distinct head pattern. White back, rump and underwing diagnostic in flight. Less gregarious than whimbrel. **Adult**: Head, neck and breast with a pale buffish-brown wash and streaked dark brown. Supercilium indistinct. Chin whitish. White lower back and rump. Tail brownish with darker streaks. White below with variable amounts of streaking on flanks and underwing. **Immature**: More buffish than adults with less streaking on breast.
Call: Characteristic .. *cur-lew* .. or .. *coor-lee* .. and also .. *tyu-yu-yu-yu-yu* ...
Range: Winter visitor. Entire seaboard.
Habitat: Beaches, tidal mudflats and estuarine swamps.

389. Blacktailed Godwit *Limosa limosa*

Identification: 41–50 cm (female larger). A tall sandy-brown and white shore bird with long slightly upturned beak. White trailing edge of wings, white rump and black-bordered white tail diagnostic in flight. **Adult nonbreeding**: Grey above and white below. White supercilium. Orange beak with black tip. **Immature**: A pale rufous-cinnamon wash on entire plumage. **Adult breeding**: Head, neck and breast rufous barred and dotted white and black increasingly on belly. Under tail-coverts white. White supercilium.
Call: Rendered as .. *wit-wit-wit* .. or .. *quick-quick-quick* .. while taking off.
Range: Winter visitor. Less in the south. The coastal and flatter sides of Southwestern India.
Habitat: Edges of lakes, tanks and estuaries.

391. Bartailed Godwit *Limosa lapponica*

Identification: 36 cm. A tall, almost straight-beaked shore bird. Smaller with shorter legs than blacktailed godwit. Grey-brown plumage on back with a scaly pattern. In flight absence of white wingbar and black on tail diagnostic. **Adult non-breeding**: Grey-brown above with pale supercilium. White below. Finely barred white tail. Bicoloured orange and black beak. **Immature**: Like non-breeding adult but with a brownish-buff wash on feathers and also with dark centres making the back appear more dotted than scaly. **Adult breeding**: Chestnut head, neck, breast and underside; the latter often with fine flecks on flanks. Back mottled grey and deep rufous. Contrasting whitish underwing diagnostic in flight.
Call: Variably rendered as .. *tak-tak-tak* .. and .. *te-ten* .. *te-ten* .. or .. *chitiu* .. *chitiu*..
Range: Winter visitor. Seaboard as far south as Mumbai.
Habitat: Beaches and tidal mudflats.

392. Spotted Redshank *Tringa erythropus* (Dusky Redshank)

Identification: 33 cm. A slim, red-legged sandpiper. Long bicoloured beak curved at tip. Plumage grey with whitish face and underside. Wings entirely mottled grey with no white bar diagnostic in flight. **Adult non-breeding**: Pale grey-brown above spotted with white. White face and supercilium contrasting against dark band through eyes. White below washed grey on sides of breast. **Immature**: Has a brownish wash over entire plumage. Chin and throat whitish. Rest of the underparts greyish with brownish barring. Base of beak red. **Adult breeding**: Sooty-black on head, neck and entire underparts except around eyes. White ring around eye, back and upper rump. Dark mantle spotted and fringed with white.
Call: Flight call .. *chu-it* .. or .. *tiu-it* .. *tiu-tiu-tiu* .. when taking off.
Range: Winter visitor along the seaboard and eastern plains of Southwestern India.
Habitat: Marshes, reservoirs and estuaries.

394. Common Redshank *Tringa totanus*

Identification: 28 cm. A slim, grey and white sandpiper with long red legs. Bicoloured beak straight and shorter than spotted redshank. In flight white back, rump and the lower base of wings (secondaries) contrast sharply against the blackish flight feathers. **Adult non-breeding**: Grey with paler ring around eyes. Underside whitewashed grey on breast and streaked and dotted with black. Base of beak red. **Immature**: Like non-breeding adult but with a warm yellowish-brown wash on plumage. Feathers on back with distinctly paler fringes giving a mottled appearance. **Adult breeding**:

Plumage variable. Characteristic blackish streaks and spots all over except on back.

Call: Characteristic .. *tewn-tewn-tewn* .. or .. *tiu-tiu-tiu* .. while taking off. Confusable with that of common greenshank.

Range: Winter visitor along the seaboard and eastern margins of Southwestern India.

Habitat: Marshes, estuaries, tidal mudflats along river mouths and salt pans.

395. Marsh Sandpiper *Tringa stagnatilis*

Identification: 25 cm. A slim, very long-legged, pale sandpiper with slender, straight black beak. White supercilium. In flight no wingbar. Long toes project well beyond whitish tail. **Adult non-breeding**: Grey above; the feathers fringed white. Wings darker. White back, rump and tail; the latter finely barred with grey. White below. Legs olive-green. **Adult breeding**: Browner plumage with heavy streaking on head and neck. Black drops or chevrons on breast. Legs yellowish. Confusable at a glance with common greenshank in all plumages except by size. Common greenshank is, however, much larger.

Call: Flight calls rendered as a shrill .. *yip* .. or .. *plew* .. or .. *che-weep* .. repeated as the bird takes off.

Range: Winter visitor along the seaboard and eastern margins of Southwestern India.

Habitat: Marshes, edges of tanks and estuaries including salt pans and tidal mudflats.

396. Common Greenshank *Tringa nebularia*

Identification: 36 cm. The largest of our sandpipers. Pallid-grey and white long-legged shore bird with long, thick, slightly upturned black beak. Overall larger size and heavier build separate from similar marsh sandpiper. **Adult non-breeding**: Grey above, paler on head and neck which are also finely streaked. White fringes to feathers on back giving it a scaly appearance. Underparts white. Greenish-yellow legs. In flight, white back visible and toes project only slightly beyond white and grey tail. **Adult breeding**: A more scaly pattern on back. Head and neck heavily streaked with black.

Call: Lower in pitch than that of redshank. .. *Tiu-tiu-tiu* .. or *tewn-tewn-tewn* ...

Range: Winter visitor along the seaboard and eastern margins of Southwestern India.

Habitat: River banks, edges of tanks, estuaries, tidal mudflats and salt pans.

397. Green Sandpiper *Tringa ochropus*

Identification: 24 cm. A small darklegged sandpiper with a typical hunched and horizontal stance at rest. Dark olive-blackish upperparts and white

underside finely streaked on neck and breast. Short white eyebrow. White rump contrasting against dark back and wings diagnostic in flight. Solitary.
Call: Flight calls .. *ti-tui* .. or .. *twee-twee-twee* .. or .. *twit-twit-twit* ...
Range: Winter visitor. Entire Southwestern India up to 2300 m (Nilgiris).
Habitat: Freshwater marshes, rain puddles, small streams and estuarine marshes.

398. Wood Sandpiper *Tringa glareola* (Spotted Sandpiper)

Identification: 21 cm. A small, slim, long-legged, olive-brown and white sandpiper. Yellow legs and white supercilium reaching the ears, diagnostic. In flight, there is less contrast between dark back and white rump than in green sandpiper. Gregarious. **Adult**: In both breeding and non-breeding, plumages more olive and distinctly spotted than green sandpiper.
Call: Flight calls .. *chip-chip-chip* .. or .. *chiff-if-iff* ...
Range: Winter visitor. Entire Southwestern India up to 2000 m.
Habitat: Marshes and mudflats along both fresh and estuarine waters.

400. Terek Sandpiper *Xenus cinereus*

Identification: 24 cm. A small, active, shortlegged sandpiper with long yellow and black upturned beak. Legs yellow. In flight white trailing edge of wings contrasting with dark back, rump and wings diagnostic. **Adult breeding**: Grey-brown above. White below with breast grey and streaked. Beak full black.
Call: Shrill .. *twit-wit-wit-wit* ...
Range: Winter visitor to the entire seaboard.
Habitat: Beaches, tidal mudflats and estuarine marshes.

401. Common Sandpiper *Tringa hypoleucos*

Identification: 21 cm. A small, short-legged olive-brown and white sandpiper. Bobbing walk and the dark sides of breast forming a contrasting band against white underside. In flight a distinct wingbar diagnostic. Solitary. **Adult**: Rather plain back, greenish short legs and longer tail extending well beyond folded wings distinguish from wood sandpiper at rest. Bobbing the tail end and low, rather erratic flight separate from other small shore birds.
Call: Flight call .. *tee-tee-tee* .. or .. *tsee-wee-wee* ... At rest .. *wheeit-wheeit* ...
Range: Winter visitor. Entire Southwestern India up to 2100 m (Nilgiris).
Habitat: River banks, hill-streams, ponds and rain puddles, tanks, lakes and reservoirs, estuaries, tidal mudflats, salt pans and beaches.

402. Ruddy Turnstone *Arenaria interpres* (Turnstone)

Identification: 22 cm. A stocky, medium-sized shore bird with short red legs and slightly upturned thick beak. In flight a striking dark and white pattern on back, wings and tail diagnostic. **Adult non-breeding**: Dark head with white chin. Reddish-brown back. Characteristic black and white pattern on breast. Underparts white. **Adult breeding**: Overall plumage coloration richer. Whiter head with black face and sides. Darker chestnut-red back. Pattern on breast more striking.
Call: Flight calls rendered as a sharp .. *chik-ik* ...
Range: Winter visitor along the entire seaboard.
Habitat: Rocky beaches and offshore rocky islands.

405. Wood Snipe *Gallinago nemoricola*

Identification: 31 cm. A dark patterned squat, long-billed marsh bird rarely seen except when flushed. Characteristic head and face pattern of pale stripes above eyes and on crown. Fully barred underside diagnostic and in flight, rounded wings and beak held pointing down are further pointers. Wooded habitats.
Call: A low .. *chok-chok* .. when flushed.
Range: Winter visitor. Western Ghats south of Mumbai including the hills of Kerala.
Habitat: Grassy marshes in hilly areas.

406. Pintail Snipe *Gallinago stenura*

Identification: 27 cm. A medium-sized, short-tailed and short-billed snipe. White belly separates from wood snipe. At rest broader supercilium and in flight indistinct pale trailing edge of wings separate from common snipe. Presence of pin-like outer tail feathers and toes noticeably projecting beyond tail are further pointers to identifying this species in flight. Prefers drier ground.
Call: Rendered variously as .. *squik* .. *chet* .. *etch* .. *scape* .. or .. *pench* .. when flushed.
Range: Winter visitor. Entire Southwestern India.
Habitat: Edges of tanks, reservoirs, flooded paddy stubble and other wetlands with short grass.

407. Swinhoe's Snipe *Gallinago megala*

Identification: 29 cm. Not readily distinguishable from pintail snipe in the field. However, Swinhoe's snipe is slightly larger, with longer tail projecting beyond tips of wings at rest. White edges to tail feathers are more distinct in flight than pintail snipe's. In hand, less tail feathers (20 as against 26 in pintail snipe) and broader pin-like outer tail feathers (2–4 mm vs 1–2 mm in pintail snipe's) are diagnostic.

Call: Similar to that of pintail snipe.
Range: Winter visitor. Reports in Southwestern India from Mumbai, south Karnataka, Kerala and Nilgiris.
Habitat: Marshes alongside pintail snipe.

409. Common snipe *Gallinago gallinago* (Fantail Snipe)

Identification: 27 cm. A pale (-) coloured snipe with white belly. Prominent white trailing edge to wings diagnostic in flight. White on belly and underwing more distinct and extensive than in pintail snipe.
Call: Similar to that of pintail snipe.
Range: Winter visitor in Southwestern India.
Habitat: Marshes.

410. Jack Snipe *Lymnocryptes minima*

Identification: 21 cm. A small narrow-winged snipe resembling common snipe. Lacks crown stripe. Supercilium double with black between the stripes. Glossy upperparts. Wedge-shaped tail. Bobs while feeding. In flight white trailing edge to wings visible. Wings darker and underwing shows less white than in common snipe.
Call: Rendered as a soft .. *gah* .. when flushed.
Range: Winter visitor in Southwestern India.
Habitat: Marshes.

411. Eurasian Woodcock *Scolopax rusticola* (Woodcock)

Identification: 36 cm. A large, heavy, long-billed woodland snipe. Camouflaged mottled brown and black plumage. Sloping forehead and eyes placed well behind. Transverse bars on crown and nape. Broad reddish wings with no distinctly paler wingbar diagnostic in flight. Flight short, rising almost vertically when flushed.
Range: Winter visitor to Southwestern India particularly in the hills from 600 to 2100 m.
Habitat: Dense evergreen forests and the associated swamps and streams.

414. Sanderling *Calidris alba*

Identification: 19 cm. A small, thickset very pale shorebird with a heavy black, short beak. Active on beaches. Dark flight feathers contrast with paler back and tail in flight. Distinct broad white wingbar diagnostic. **Adult non-breeding**: Pearly-grey above. White below. **Adult breeding**: Chestnut head, neck, back and upper breast heavily mottled and streaked with brown and white. **Immature**: Like non-breeding adults but with mottled back..
Call: Flight calls rendered as .. *twick* .. or .. *wick-wick* ...

Range: Winter visitor along the seaboard. Less in the south.
Habitat: Sandy beaches.

416. Little Stint *Calidris minuta*

Identification: 15 cm. Our smallest shorebird (slightly larger than the house sparrow). Mottled greyish-brown and white with short black beak and legs. In flight white wingbar visible and smoky-grey outer tail feathers diagnostic. Large flocks. **Adult breeding** and **Immature**: Show more rufous on upper plumage. Head entirely rufous in adults as against grey with a whitish supercilium in immature.
Call: Flight calls rendered as .. *wit-wit-wit* .. and .. *trrr* ...
Range: Winter visitor along the seaboard.
Habitat: Beaches, tidal mudflats, salt pans and marshes.

417. Temminck's Stint *Calidris temminckii*

Identification: 15 cm. Small and confusable with little stint. Overall appearance greyer and patterned like common sandpiper. Yellowish legs and white outer tail feathers (in flight) separate from little stint. **Adult non-breeding**: Never shows a paler supercilium. More grey on breast than little stint. Tail projects well beyond closed wing-tips at rest. **Adult breeding**: Pale rufous wash on entire upper plumage including head and breast. **Immature**: Intermediate between breeding and non-breeding adults. Rufous touch on back. Head, neck and breast greyish.
Call: Flight note rendered as .. *tiriririr* .. different from that of little stint.
Range: Winter visitor along the seaboard and the eastern margins of Southwestern India.
Habitat: Freshwater marshes, tidal mudflats, estuaries and beaches.

420. Dunlin *Calidris alpina*

Identification: 19 cm. A typically hunched sandpiper with black legs and long, slightly decurved black beak. White wingbar and white sides to rump and tail diagnostic. **Adult non-breeding**: Grey above including head, neck and breast. Chin, lower breast and underparts white. Lacks a distinct supercilium (*Compare with* Curlew Sandpiper, 422). **Adult breeding** and **Immature**: Rufous, white and black mottled pattern on mantle. Black streaks on head, neck and breast. Immature birds lack the black belly of breeding adults.
Call: Flight call rendered as .. *teu-eep* .. or .. *wee-wee-eet* .. or .. *treeep* ...
Range: Winter visitor as far south as Mumbai along the western seaboard. Stragglers noticed in Uttara Kannada district (Karnataka) and Kerala.
Habitat: Beaches, tidal mudflats and salt pans.

422. Curlew Sandpiper *Calidris ferruginea*

Identification: 20 cm. A small shorebird resembling dunlin. Longer, decurved beak and a prominent pale supercilium distinguish from dunlin at rest. In flight full white rump diagnostic. **Adult non-breeding** and **Immature**: Greyish above and white below. Immature birds have the mantle browner with a distinct scaly pattern. **Adult breeding**: Chestnut head, neck, breast and underparts. Mantle darker with a bright scaly pattern. Contrasting white underwing diagnostic in flight.

Call: Rendered as .. *chirrup* ..

Range: Winter visitor along the seaboard.

Habitat: Beaches, tidal mudflats, estuarine marshes and salt pans.

424. Broadbilled Sandpiper *Limicola falcinellus*

Identification: 17 cm. A shorebird resembling a large stint with a long beak slightly bent at the tip. Double supercilium diagnostic in àll plumages. In flight strikingly dark wings distinctly paler in the centre and edged with a white bar visible. **Adult non-breeding**: Upper supercilium not very prominent. Dark shoulders prominent on folded wings. **Adult breeding** and **Immature**: Browner mantle heavily streaked with black and white. Head, neck and breast streaked with black. Blackish rump and white sides diagnostic in flight.

Call: Flight call rendered as .. *chrrreet* ... Others described as similar to Temminck's stint .. *tzit* ...

Range: Winter visitor along the seaboard.

Habitat: Beaches, tidal mudflats and estuarine marshes.

426. Ruff *Philomachus pugnax*

Identification: 25–30 cm. A long-legged grey and white shorebird with red legs confusable with common redshank. However, overall body proportions are different: small head, slender neck, deep belly and shorter beak. In flight narrow wingbar and white sides to dark rump and tail diagnostic. **Adult Female** and **Immature**: Browner plumage with a scaly pattern. White lower breast, belly and under tail-coverts. Immature birds have greenish legs. **Breeding male**: Colour variable. Larger size. Bare face and beak reddish. Head tufts and neck plumes resemble a mane. Colour varies from almost white to full black.

Call: Flight calls rendered as .. *tu-whit* .. or ... *hoo-ee* .. A low .. *kuk-uk* .. or .. *chuck-chuck* ...

Range: Winter visitor along the seaboard.

Habitat: Mudflats, salt marshes and wet paddy stubbles.

Family ROSTRATULIDAE

Snipe-like birds. Plump and long-billed. Tailless appearance. Bob heads up and down. Flight slower than snipes. Female more colourful than male. Secretive. Mostly nocturnal.

429. Greater Painted Snipe *Rostratula benghalensis* (Painted Snipe)

Identification: 25 cm. A strikingly patterned snipe with distinctly pale spectacles around eyes and a white band from breast across shoulder and back. White underwing and black wingbar diagnostic in flight. **Adult Female**: Richly coloured plumage; maroon-brown head, neck and upper breast. White spectacles. Glossy green wings. White below. **Adult Male and Immature**: Smaller and duller than female. Upper plumage mottled and patterned with golden-buff.
Call: Female at the time of breeding .. *ook* ... Flight call rendered as .. *kek* ...
Range: Resident with local movements all over Southwestern India.
Habitat: Marshes with dense reeds, shrubs and sewage pools.

Family RECURVIROSTRIDAE

Long-legged black and white shorebirds with long beaks: straight in **Stilts** and upturned in **Avocets**.

430. Blackwinged Stilt *Himantopus himantopus*

Identification: 25 cm. Unmistakable black and white birds with long red legs. Long legs projecting beyond white tail diagnostic in flight. **Adult Female and Immature**: Black on wings duller and browner. Brownish on crown, head and hindneck.
Call: Flight call .. *kip* .. *kip* ...
Range: Resident with local movements along the margins of Southwestern India.
Habitat: Marshes including estuaries, mangroves and salt pans.

432. Pied Avocet *Recurvirostra avosetta* (Avocet)

Identification: 46 cm. White with black head and hindneck. Black upturned beak diagnostic. Black bands on shoulders and black flight feathers contrasting with white body readily identify avocets at rest and in flight. Blue-grey legs.
Call: .. *Klooit* ...
Range: Winter visitor along the seaboard as far south as northern Karnataka.
Habitat: Marshes, mostly of saltwater including salt pans.

Family DROMADIDAE

Stocky shorebirds with short, heavy, dagger-like beaks. Head and neck held out in flight with the long legs trailing behind.

434. Crab Plover *Dromas ardeola*

Identification: 41 cm. Overall plumage resembles pied avocet. White head and thick, short, black beak diagnostic. White body with black back and black flight feathers are pointers in flight. **Immature**: Black on plumage replaced by silvery-grey. Head with black streaking on rear crown.
Call: Rendered as .. *ka* .. and .. *kwe-ki-ki* .. (near nest).
Range: Sparse winter visitor along the seaboard.
Habitat: Rocky beaches and tidal mudflats.

Family BURHINIDAE

Large shorebirds with short thick beaks. Large eyes. Long legs with only 3 toes. Nocturnal.

436. Eurasian Thick-knee *Burhinus oedicnemus* (Stone Curlew)

Identification: 41 cm. An earthy-brown bird with neatly camouflaging plumage. Long yellow legs with thick knees. Big yellow eyes and short yellow beak with a black tip. Alternating black and white bands on closed wings. In flight, black flight feathers with white mirrors diagnostic.
Call: Named after its call that resembles the curlew. Rendered as .. *pick* .. *pick* .. *pick* .. *pick* .. *pick-wick* .. *pick-wick* .. *pick-wick* ...
Range: Resident with local movements. Entire Southwestern India up to 1000 m.
Habitat: Stony semi-desert and dry cultivation.

437. Great Thick-knee *Burhinus recurvirostris* (Great Stone Plover)

Identification: 51 cm. A very large greyish shorebird with unstreaked plumage, bold black and white facial pattern, large yellow and black up-curved beak and long yellow legs. Found near water. In flight, rounded wings, shorter tail (duck-like), unmarked dorsal plumage and larger white mirrors on black flight feathers separate from stone curlew.
Call: Rendered as a loud .. *kree-kree-kree* .. *kre-kre-kre-kre* ...
Range: Resident with local movements. Entire Southwestern India.
Habitat: Rocky riverbeds, lake margins, estuaries and beaches.

Family GLAREOLIDAE

Sandy-brown birds with short, slightly hooked beaks. **Coursers**: long-legged ground birds resembling the lapwings, though more upright in stance. Run fast on open ground. **Pratincoles**: tern-like shorebirds with short legs and long, slender wings. Fly gregariously near water, hawking insects on wings.

440. Indian Courser *Cursorius coromandelicus*

Identification: 26 cm. A sandy-brown fast-running long-legged ground bird with whitish legs. Chestnut cap, black line through eyes and white supercilium. Full black flight feathers and white rump diagnostic in flight.
Range: Resident with local movements along the drier eastern margins of Southwestern India straggling westward to the coast.
Habitat: Stony plains, grassland and dry cultivation.

443. Oriental Pratincole *Glareola maldivarum* (Large Pratincole)

Identification: 24 cm. A brown, long-winged tern-like bird that feeds aerially. Uniform dusky-brown upperparts, short black deeply forked tail and white rump diagnostic in flight. **Adult breeding**: Black line from below eyes forms a neat necklace below pale throat. Base of beak red. **Non-breeding** and **Immature**: Whitish throat and abdomen. Black necklace obscure or absent.
Call: Flight call rendered as .. *kirri-kirri* .. or .. *chik-chik* ...
Range: Winter stragglers in Southwestern India.
Habitat: Marshes and rivers.

444. Small Pratincole *Glareola lactea*

Identification: 17 cm. Small swallow-like greyish bird. Pale back and wings with black flight feathers and black tips to short white tail diagnostic in flight. Broad white trailing edge of wing with black border visible from above and below when the bird flies.
Call: Rendered as .. *tuck-tuck-tuck* .. and also as .. *pirrip* .. or .. *tirrit* ...
Range: Resident with local movements over Southwestern India.
Habitat: Lakes, reservoirs, rivers, estuaries and beaches.

Family STERCORARIIDAE

Large predators of open ocean and coast. Long wings and fast direct falcon-like flight. Tails often with elongated central feathers. Feed among gulls frequently as pirates. Feet webbed. Rest on water.

446. Brown Skua *Cartharacta lonnbergi* (Antarctic Skua)

Identification: 53–61 cm. Large. Uniform brown with black flight feathers. White patch on primaries conspicuous. At closer range, spots and streaks on plumage and darker cap visible.
Range: Straggler to the coasts of Southwestern India recorded from Kerala, Karnataka and Ratnagiri district (Maharashtra).
Habitat: Marine.

447. Pomarine Jaeger *Stercorarius pomarinus* (Pomatorhine Skua)

Identification: 53 cm. Occurs in both dark and pale plumages. Long spoon-like and twisted central tail feathers diagnostic. In winter when the tail lacks the long feathers, broad bands on upper and under tail-coverts are pointers.
Range: Straggler to the southwestern coast (Mumbai).
Habitat: Marine.

448. Parasitic Jaeger *Stercorarius parasiticus* (Parasitic Skua)

Identification: 48 cm. A small skua often seen chasing gulls over the waters. Colour variable; pale or dark. Tail streamers are stiff, long and dagger-like. In winter, lacks the streamers. Less barring on tail-coverts diagnostic.
Range: Stragglers along the southwestern coasts. Recorded from Goa, Uttara Kannada (Karnataka) and Mumbai.
Habitat: Marine, often close to the shore.

Family LARIDAE

Seabirds with largely white and grey plumage, thick dagger-like beaks and webbed feet. **Gulls**: Larger and heavier in build, broader wings and short square tail. Float on water. Scavenge along the shore. **Terns**: Slender, long winged birds often with long deeply forked tails. Plunge feed in water. Rarely rest on water.

449. Sooty Gull *Larus hemprichii*

Identification: 48 cm. A medium-sized gull with a dark hood and brownish upperparts. A white trailing edge to wings may be present in flight. No white wing mirrors. Yellow beak and legs, the former tipped black and yellow, diagnostic.
Range: Winter straggler to Mumbai.
Habitat: Marine.

450, 451. Yellowlegged Gull *Larus cachinnans* (Herring Gull)

Identification: 60 cm. Large. White with silvery-grey wings. In flight black wing tips with white mirrors and white trailing edges to wings diagnostic. Yellow beak and legs, the former tipped red. **Adult non-breeding** and **Immature**: Plumage on head streaked with grey. Immature is fully mottled with black beak.
Call: Rendered as .. *kee-ow* .. *kee-ow* ...
Range: Winter visitor along the seaboard. Reported from Mumbai and Kerala.
Habitat: Marine.

452. Lesser Blackbacked Gull *Larus fuscus*

Identification: 60 cm. Very similar to herring gull. However, mantle is blackish and in flight contrasts strikingly against rest of the plumage which is white.
Call: Rendered as similar to herring gull's but louder.
Range: Winter visitor along the seaboard.
Habitat: Marine. Along beaches, estuaries and salt marshes including salt pans.

453. Great Blackheaded Gull *Larus icthyaetus*

Identification: 66–72 cm. Largest of our gulls. Full black head (breeding). Yellow beak with black band near tip. Slow flapping flight. In flight black subterminal band on white tail diagnostic and separates from yellowlegged gull.
Call: Loud .. *kraaa* ...
Range: Winter visitor along the seaboard.
Habitat: Marine.

454. Brownheaded Gull *Larus brunnicephalus*

Identification: 46 cm. Medium-sized gull with blackish-brown head (breeding). Bright red beak and legs, the former dark-tipped. Black wing-tips with two large white mirrors diagnostic. **Adult non-breeding**: White head with a black crescent-shaped mark behind ear. **Immature**: Lacks wing-mirrors. More mottled plumage. Black subterminal band on tail. Broader black trailing edge to wings separates from similar immature common blackheaded gull.
Call: A loud .. *kreeak* ...
Range: Winter visitor along the entire seaboard.
Habitat: Beaches, estuaries and salt marshes including salt pans. Occasionally inland in lakes and rivers.

455. Common Blackheaded Gull *Larus ridibundus*

Identification: 43 cm. In all plumages similar to brownheaded gull except
the wing which is largely white with black tips and no mirrors. At rest
smaller size can help separate this from the larger species. **Immature**:
Resemble immature brownheaded gull except in the amount of black on
the flight feathers. Tail with the black band.
Call: Loud .. *kree-ah* .. and .. *ka-yek* ...
Range: Winter visitor along the seaboard.
Habitat: Marine.

456. Slenderbilled Gull *Larus genei*

Identification: 43 cm. Confusingly similar to common blackheaded gull in
overall plumage and size. However, head is white in all plumages and may
have an indistinct dark spot behind ears in winter. Legs and beak scarlet.
White iris. **Non-breeding** and **Immature**: Pale orange legs and beak.
Longer neck and beak with a sloping forehead at rest and in flight, long-
necked, long-tailed and humpbacked profile separate from common blackheaded gull.
Range: Winter visitor to the coasts of Gujarat and Mumbai. Stragglers (?)
recorded from the coasts of Uttara Kannada (Karnataka).
Habitat: Marine. Also seen in estuaries and salt pans.

458. Whiskered Tern *Chlidonias hybridus*

Identification: 25 cm. Small. White and grey tern with black beak and feet.
Tail short and slightly forked. Black patches from eyes extend to hindcollar.
At rest closed wing-tips project beyond tail. **Adult breeding**: Black cap
and stripe across cheeks. Dark grey underparts. Beak and legs dark blood-
red. **Immature**: Browner upperparts with greyish rump and wings. Head
dark brown.
Call: Harsh .. *kreak* .. *kreak* ...
Range: Winter visitor in Southwestern India.
Habitat: Lakes, marshes, estuaries and tidal mudflats.

459. Whitewinged Tern *Chlidonias leucopterus* (Whitewinged Black Tern)

Identification: 23 cm. Very similar to whiskered tern except in having a
shorter beak, white hindcollar and whiter rump. In partial summer it shows
contrasting dark and white plumage; white forehead and tail and dark grey
underparts including underwing. **Adult breeding**: Blackish head, body and
underwing contrasting with white wings, tail and under tail-coverts diagnostic.
Call: Similar to that of whiskered tern.

Range: Winter visitor straggling along the western seaboard. Recorded from Gujarat and Mumbai.
Habitat: None recorded specifically for western India. Prefers freshwater marshes and rivers.

460. Gullbilled Tern *Sterna nilotica*

Identification: 38 cm. A white and grey tern larger than whiskered tern with thick black beak and legs and more deeply forked tail. Black streaked head with black patch around eyes and over ears. **Adult breeding**: Black cap. Lack of crest, full black beak and overall stockier build separate from sandwich tern.
Call: Rendered as .. *wik... kuwikkeewik* ..
Range: Winter visitor to Southwestern India.
Habitat: Lakes, rivers, estuaries, beaches and salt pans.

462. Caspian Tern *Sterna caspia*

Identification: 51 cm. The largest of our terns. Large size with red beak and black legs diagnostic. **Adult non-breeding** and **Immature**: Pearl-grey above (streaked in immature) with head and neck white. Black streaks on cap. **Adult breeding**: Black crown, wing-tips and legs contrasting against pale body and bright red large bill (tipped darker) diagnostic in flight.
Call: Rendered as a loud .. *krake-kra* ...
Range: Winter visitor along the seaboard.
Habitat: Marine including estuaries. Occasionally on inland lakes.

463. River Tern *Sterna aurantia*

Identification: 38–46 cm. A medium-sized slender, graceful grey and white tern with blackish crown (full black at breeding), yellow beak and red legs. Deeply forked swallow-like long white tail. **Immature**: Mottled brownish mantle. Differs from blackbellied tern at all ages by having fully white underparts.
Call: A shrill .. *kreek* .. often repeated. Also rendered as .. *ping* ...
Range: Resident with local movements. All over Southwestern India.
Habitat: Larger reservoirs, lakes and rivers. Occasionally estuaries.

465. Common Tern *Sterna hirundo*

Identification: 36 cm. A medium-sized grey and white tern with white forehead and black-streaked cap (full black in breeding). Red beak tipped black. Red legs. Long, deeply forked grey and white tail does not project beyond wing-tips at rest. **Adult breeding** and **Immature**: Confusable with roseate tern. However, adult common tern has darker underparts at

breeding. **Immature**: roseate tern has darker cap and back and whiter underwing. Shorter tail of common tern diagnostic at all plumages.
Range: Rare winter visitor over the seaboard.
Habitat: Rivers and estuaries.

466. Roseate Tern *Sterna dougallii*

Identification: 38 cm. A medium-sized tern, barely distinguishable from common tern except by longer tail (projecting beyond wing-tips at rest) which is full white. **Adult breeding**: White underparts with a pinkish flush. (For separating immature birds *see* under Common Tern, 465)
Range: Resident with local movements along the coastal islands. Breeds on the Vengurla Rocks off the coast of Maharashtra.
Habitat: Marine.

467. Whitecheeked Tern *Sterna repressa*

Identification: 35 cm. A medium-sized tern confusingly similar to common tern and roseate tern. Obvious grey rump and tail and darker upperwings contrasting with silvery bases diagnostic. **Adult breeding**: White cheek separating black crown from dark underparts.
Range: Resident with local movements along the western coast. Breeds on the Vengurla Rocks. Specimens from Mumbai.
Habitat: Marine.

470. Blackbellied Tern *Sterna acuticauda*

Identification: 33 cm. A medium-sized freshwater tern rather similar to river tern. Black belly and deeply forked long tail diagnostic. **Adult breeding**: Black cap with bright yellow beak and orange-red legs. Uniform grey upperparts. Black belly patch contrasting with white underparts. **Adult non-breeding**: Beak tipped black. Crown and belly with traces of black; the latter occasionally fully white. Separated from river tern by smaller size, slenderer beak, lighter flight and longer tail. **Immature**: Mottled buffish above.
Call: Rendered as a shrill .. *krek* .. *krek* ...
Range: Resident. Widespread along the eastern parts of Southwestern India.
Habitat: Freshwater lakes and inland rivers.

472. Bridled Tern *Sterna anaethetus* (Brownwinged Tern)

Identification: 37 cm. A medium-sized black-capped tern with white forehead and supercilium. Blackish upperparts and white underparts diagnostic. Long tail deeply forked. Sooty tern is darker. **Adult non-breeding**: White speckles on crown.

Call: Rendered as .. *krek* .. or .. *quirk* ...

Range: Resident with local movements along the seaboard. Breeds on the Vengurla Rocks (Maharashtra).

Habitat: Marine.

474. Sooty Tern *Sterna fuscata*

Identification: 43 cm. A medium-sized tern slightly larger than bridled tern. Darker sooty-grey-blackish. White restricted to forehead and not beyond eyes. Broader blackish trailing edge to wings in flight. **Immature:** Differs from bridled tern in having wholly brown head and upper breast. Rest of the underparts white.

Call: Rendered as .. *wide-wake* .. hence popularly known as 'Wideawake Tern'.

Range: Stragglers along the coastline. Breeds on Vengurla Rocks (Maharashtra).

Habitat: Marine.

475, 477. Little Tern *Sterna albifrons*

Identification: 23 cm. A small grey and white tern. Blackish beak and dusky-red legs. Black cap mixed with white. Difficult to separate from Saunders' tern except at breeding. **Adult breeding:** Black cap with white forehead (extending over eyes as supercilium). Beak and feet orange-yellow. **Immature:** Dark wavy bars on upperparts. Forehead and crown white speckled with brown. Races separable only in hand. Shafts of first three primaries brown in race *albifrons* and white in race *pusilla*.

Call: Rendered as a shrill .. *creek* .. *creek* ...

Range: Resident along the coasts of northern Gujarat and offshore Mumbai. Winter stragglers over the seaboard as far south as Kerala.

Habitat: Freshwater marshes, rivers and estuaries.

476. Saunders' Tern *Sterna saundersi*

Identification: 23 cm. Very similar to little tern. Legs and beak blackish. Forehead and crown whiter than in little tern. **Adult breeding:** Black cap. Forehead white. Black beak tipped yellow. Legs brownish. Darker outer primaries and white forehead not extending over eyes separate from breeding little tern.

Range: Stragglers collected from the coasts of Gujarat and Mumbai.

Habitat : Sandy beaches, offshore islets and estuaries.

478. Great Crested Tern *Sterna bergii* (Large Crested Tern)

Identification: 53 cm. A large tern with yellow beak and black legs. A prominent black crest from nape. **Adult breeding:** Black cap and crest.

Narrow white forehead diagnostic. **Immature**: Like non-breeding adults. Back more mottled brownish. Yellow beak diagnostic in all plumages. **Call**: Rendered as .. *chirruk* ... **Range**: Breeds on offshore islands in the Arabian Sea. Non-breeding birds wander all over the seaboard. **Habitat**: Open sea, beaches and rocky islets close to the shore.

479. Lesser Crested Tern *Sterna bengalensis*

Identification: 43 cm. Medium-sized tern with black crest and orange beak. Broad white forehead diagnostic. **Adult breeding**: Full black crown without white forehead diagnostic. **Immature**: Smaller size and orange beak separate from immature great crested tern. **Call**: Similar to that of great crested tern. **Range**: Presumably resident. Seen all through the year along the seaboard. **Habitat**: Open sea and beaches.

480. Sandwich Tern *Sterna sandvicensis*

Identification: 44 cm. Medium-sized pallid tern confusable with gullbilled tern. Black crest on nape, deeper fork on tail, overall slender build and longer black beak tipped yellow diagnostic. **Adult breeding**: Black crown and crest. Pale plumage and dark outer primaries diagnostic in flight. **Immature**: Like non-breeding adult but with dark band on wings and tail. Primaries dark grey in flight. **Range**: Winter visitor to the seaboard. Gujarat to northern Karnataka (Karwar). **Habitat**: Marine.

484. Indian Skimmer *Rynchops albicollis*

Identification: 40 cm. An unmistakable medium-sized black and white tern-like bird with long red beak and red legs. **Adult**: Beak compressed and knife-like. The lower mandible longer than upper. Forehead white. **Immature**: Brownish above with duskier beak. Head whitish. **Call**: Rendered as a nasal .. *kap* .. *kap* ... **Range**: Stragglers recorded from Kutch, Mumbai and northern Karnataka (Gokarna). **Habitat**: River mouths and estuaries.

Family PTEROCLIDIDAE

Short-legged pigeon-like terrestrial birds capable of powerful flight and fast walking. Partridge-like colour pattern camouflages the birds well against the earthy background in their habitats.

487. Chestnutbellied Sandgrouse *Pterocles exustus* (Indian Sandgrouse)

Identification: 28 cm. Yellowish sandy-grey pigeon-like birds with a narrow black band across breast. Stout body, slender bow-like wings and tapering tail diagnostic. **Adult Male**: Upperparts with crescentic marks and speckles. Cheeks, chin and throat dull yellow. Chocolate-blackish belly. **Adult Female**: Dull buff and brown upperparts with a mottled pattern as in male. Pale underparts spotted on upper breast and barred blackish on abdomen and flanks. **Immature**: Upperparts as in adult female but finely vermiculated with black. Underparts finely barred on chin, throat and breast. Abdomen dull blackish.
Call: Rendered as *.. gutu gutu ...* or *.. waku-waku ...* In flight *.. kut-ro ...*
Range: Resident with local movements. Entire Southwestern India along the drier eastern margins.
Habitat: Rocky scrub, dry cultivation and riverbeds.

Family COLUMBIDAE

Plump and round-bodied with small head and beak. Short legs. Longish pointed wings. Powerful flight, flapping loudly while taking off. **Imperial, Green** and **Wood Pigeons** are largely arboreal and coloured green or maroon. The **Rock Pigeon** is the common pigeon which is blue-grey. **Doves** are terrestrial, slenderer and usually dull coloured (exceptionally green) with spots and vermiculations on neck and back.

496. Pompadour Green Pigeon *Treron pompadora*
(Greyfronted Green Pigeon)

Identification: 28 cm. A small stocky green pigeon with red legs. Maroon-chestnut back identifies the males easily. **Adult Male**: Forehead, crown and nape ashy-grey. Back and scapulars maroon-chestnut. Black shoulder. Broad yellow wing band. Middle pair of tail feathers olive-green. Yellowish-green below; upper breast tinted with orange or pinkish. Cinnamon under tail-coverts. **Adult Female**: Overall olive-green lacking the maroon back and pinkish breast of males. Green and white under tail-coverts. Olive-green middle tail feathers separate from similar female orange-breasted green pigeon.
Call: A mellow whistle *.. phoi-phoi-oi-oi-oi ...*
Range: Resident with local movements. Western Ghats up to 1200 m.
Habitat: Evergreen and moist deciduous forests including their stages of degradation, teak plantations and within small towns.

501. Orangebreasted Green Pigeon *Treron bicincta*

Identification: 29 cm. Very similar to pompadour green pigeon. Tail greyish-slaty with a blackish subterminal band except on the middle pair which is fully grey. **Adult Male**: Greenish head and back. Distinct pinkish and orange bands across breast diagnostic. Rest of the plumage as in pompadour green pigeon except tail. Grey tail is unmistakable. **Adult Female**: Separated from female pompadour green pigeon by full grey middle tail feathers.
Call: Described as hoarser or lower and more jerky than pompadour green pigeon's.
Range: Resident with local movements. Western Ghats. Belgaum southwards, locally absent or rare. Found in the hills up to 1000 m.
Habitat: Evergreen and moist deciduous forests.

504. Yellowfooted Green Pigeon *Treron phoenicoptera*
(Common Green Pigeon)

Identification: 33 cm. Yellow legs. Larger than pompadour green pigeon and orangebreasted green pigeon. Greyish and olive-green above. Yellowish below. Pinkish shoulder patch. Sexes more or less similar.
Call: Rather like pompadour green pigeon's but more jerky and lower in tone.
Range: Resident with local movements. Entire Western Ghats.
Habitat: Deciduous and open secondary forests as well as outskirts of towns.

507. Green Imperial Pigeon *Ducula aenea*

Identification: 43 cm. Large size. Pinkish-grey body with metallic bronzy-green back, wings and tail. Chestnut under tail-coverts distinctly darker than belly and tail. Characteristic rolling display flight. Sexes alike.
Call: Loud cooing rendered as .. *click-hrooo* .. or .. *wuck-wuck-woor* .. and .. *wuck-woor-woor-woor* ...
Range: Resident with local movements. Western Ghats south of Mumbai up to 600 m.
Habitat: Deciduous and open, moist secondary forests, mostly being absent in the wetter coastal evergreen forests.

511. Mountain Imperial Pigeon *Ducula badia*
(Maroonbacked Imperial Pigeon)

Identification: 43 cm. Large. Pale white-grey forest pigeon with brownish-slate and maroon back and wings. Dark tail with a distinctly paler broad terminal band. Flight silhouette much like a largebilled crow at a distance. Sexes alike.
Call: Loud roaring .. *uk-ook* .. *uk-ook* .. or .. *groo-groo* ...

Range: Resident with local movements. Western Ghats south of Belgaum, from the coast to about 2000 m.

Habitat: Dense evergreen and montane forests and secondary forests in the wetter slopes of the Ghats.

517. Rock Pigeon *Columba livia* (Blue Rock Pigeon)

Identification: 33 cm. An unmistakable bluish-grey pigeon with glistening purplish and green neck, shoulders and breast. Two bars on wings. Sexes alike.

Call: A deep .. *gootr-goo* .. and .. *goo-goo-goo* ...

Range: Resident. Entire Southwestern India up to over 2000 m.

Habitat: Cliffs, waterfalls, gorges, dams, bridges and buildings from suburbs to the middle of dense evergreen forests.

521. Nilgiri Wood Pigeon *Columba elphinstonii*

Identification: 42 cm. Large. Dark metallic greenish-purple-brown forest pigeon with grey underparts. Grey underwing and undertail. Pale head with a distinct black and white checkerboard collar on hindneck diagnostic. Sexes alike.

Call: A loud langur-like .. *boom* ... Also series of hoots .. *who-who-who* ...

Range: Endemic. Resident with local movements. Entire Western Ghats south of Mumbai up to over 2000 m. Locally rare or absent. An isolated population has been reported from Nandi hills (Bangalore) much to the east of the range of the species.

Habitat: Evergreen and montane forests, secondary moist forests and monocultures of pine (Nilgiris).

531, 532, 533. Oriental Turtle Dove *Streptopelia orientalis* (Rufous Turtle Dove)

Identification: 33 cm. A large reddish-brown dove with a scaly patterned back, black and white checkerboard on sides of neck and white-tipped grey tail (conspicuous in flight). Differs from commoner spotted dove in larger size and stockier pigeon-like build, neck pattern, slaty-grey lower back and rump, narrower white tips to tail, and greyish flanks and under tail-coverts. Sexes alike. The races differ in the colour of head, under tail-coverts and tip of tail. Race *erythrocephala* has a reddish head. Races *meena* and *agricola* have grey heads. White under tail-coverts and tail tips identify race *meena*. The corresponding colour in race *agricola* is grey.

Call: Rendered as a mournful .. *goor-gur-grugroo* ..

Range: Status in Southwestern India uncertain as the races mix in winter and the resident race *erythrocephala* is locally migratory. Not recorded south of Karnataka.

Habitat: Along edges of evergreen forests (often on the roadside), open forests, dry bamboo facies.and cultivation.

534. Eurasian Collared Dove *Streptopelia decaocta* (Ring Dove)

Identification: 32 cm. A stocky grey and brown dove with a prominent narrow black half-collar on hindneck. Underside pinkish on breast turning grey on abdomen and under tail-coverts. Blackish-grey tail with broad white tips. Sexes alike. Confusable with female red collared dove. The Eurasian collared dove is, however, larger, longer-tailed and with a conspicuous yellow ring around eye. Red legs.

Call: Rendered as a deep, pleasant .. *kukkoo* ... *kook*.. Also .. *koon* .. *koon* .. in flight.

Range: Resident with local movements along the eastern margins of Southwestern India up to 1000 m.

Habitat: Dry scrub and thicket, open deciduous forest and cultivation.

535. Red Collared Dove *Streptopelia tranquebarica* (Red Turtle Dove)

Identification: 23 cm. Small size, short tail and plump build diagnostic. Ring on neck. Outer tail feathers broadly tipped white and wing lining dark grey. Brownish legs. **Adult Male**: Grey head and throat. Brick-red-pinkish body. Slaty rump and upper tail-coverts. **Adult Female**: Duller than male. Similar to Eurasian collared dove except smaller size and shorter tail.

Call: Rendered as .. *groo-gurr-goo* .. *groo-gurr-goo* ...

Range: Status Southwestern India uncertain due the local movements of the species. Recorded in the south as far as Karnataka.

Habitat: Dry scrub and thicket as well as cultivation.

537. Spotted Dove *Streptopelia chinensis*

Identification: 30 cm. A slender, long-tailed dove with pinkish-brown body and spotted grey-brown wings and back. Dark flight feathers and broadly white-tipped dark tail conspicuous in flight. A black and white checkerboard-like half collar across hindneck diagnostic. Sexes alike.

Call: A pleasant mournful .. *kro-kro-kroo* .. *kro-kro-kro* ...

Range: Resident. Entire Southwestern India up to 1500 m.

Habitat: Edges of evergreen forests, secondary open and deciduous forests, scrub, cultivation, margins of marshes, monocultures including eucalyptus, urban and suburban gardens and buildings.

541. Laughing Dove *Streptopelia senegalensis* (Little Brown Dove)

Identification: 27 cm. Small pinkish-brown and grey dove with a small black and rufous checkerboard pattern on sides of neck. Dark wings and broadly

white-tipped dark tail conspicuous in flight. **Adult**: Grey on wing-shoulders. White abdomen and under tail-coverts. **Immature**: Lacks the checkerboard on sides of neck.
Call: A low .. *coo-rooroo-rooroo* ...
Range: Resident with local movements. Entire Southwestern India up to 1500 m.
Habitat: Dry scrub and thicket with thorn and *Euphorbia*, rocky plains, hillsides and cultivation.

542. Emerald Dove *Chalcophaps indica*

Identification: 27 cm. A small brownish-pink forest dove with bright bronzy-green wings. White supercilium on blue-grey crown and neck. Red beak and feet. In flight, grey rump and brown tail with a black band diagnostic. Sexes alike.
Call: Single low .. *hoon* ...
Range: Resident. Western Ghats up to over 1800 m.
Habitat: Evergreen forests particularly along the edges and paths, moist deciduous and secondary forests, monocultures of teak and mixed orchards of arecanut, banana and cacao in the hills. Occasionally in the backyards of homes.

Family PSITTACIDAE

Small to large grass-green birds with heavy, hooked red beaks. Slender, long wings and fast flight. Flocks. Sexes different. **Parakeets** are larger with long tails. Males have a ring around neck. **Lorikeets** are small and stump-tailed. Sleep hanging upside down.

545, 546. Alexandrine Parakeet *Psittacula eupatria* (Large Indian Parakeet)

Identification: 51 cm. A large grass-green parakeet with heavy red beak and red shoulder patch. **Male**: throat black with a rose ring running around hindneck. The races are separable only in hand. The black streak over throat is broader in race *nipalensis* then in race *eupatoria*.
Call: Harsh grating .. *keeak* ...
Range: Resident with local movements. Distributed widely on the Western Ghats up to 900 m. Race *nipalensis* occurs in Maharashtra and Gujarat. Locally rare or extinct. No recent reports from Kerala and around Jog Falls (Karnataka) where the species was once known.
Habitat: Deciduous forests, secondary moist forests and thickets interspersed with taller trees.

550. Roseringed Parakeet *Psittacula krameri*

Identification: 42 cm. A medium-sized parakeet. Grass-green with red beak. Males have a black and rose collar around hindneck. Both sexes distinguished from large Indian parakeet by smaller size, lack of red shoulder patch and black lower mandible.
Call: Shrill .. *kee-ak* ...
Range: Resident. Entire Southwestern India.
Habitat: Open forests, scrub and thicket, cultivation, orchards and monocultures, beach and estuarine vegetation including groves of coconut and casuarina, urban and suburban gardens.

558. Plumheaded Parakeet *Psittacula cyanocephala*
(Blossomheaded Parakeet)

Identification: 36 cm. Slender green parakeet with plum-red (males) or slaty-grey (females) head and whitish beak. Dashing, erratic flight in flocks.
Call: Shrill .. *tooi* .. or .. *tuee* .. in flight and at rest.
Range: Resident with local movements. Western Ghats up to 1300 m.
Habitat: Evergreen, deciduous and secondary forests, monocultures, hillside ‚ cultivation and orchards, moist scrub and thickets, coastal vegetation, urban and suburban gardens.

564. Malabar Parakeet *Psittacula columboides* (Bluewinged Parakeet)

Identification: 38 cm. Slender grey-green-whitish parakeet with blue wings, lower back, rump and tail. **Adult Male**: Whitish with blue-green and black collar. Upper mandible red. **Adult Female**: Overall greenish without the collar. Both mandibles grey-black diagnostic.
Call: Harsh loud .. *eeaah* .. uttered on wings.
Range: Endemic. Resident with local movements, especially during the rains. Between Thane district (Maharashtra) and the southern tip of the Western Ghats from the coast to over 1000 m.
Habitat: Evergreen and moist deciduous forests including their stages of degradation, moist scrub and thickets, monocultures, estuarine vegetation and suburbs.

566. Vernal Hanging Parrot *Loriculus vernalis* (Indian Lorikeet)

Identification: 14 cm. A tiny grass-green stump-tailed parrot with crimson-red beak and rump. **Male**: has a small blue patch on throat. Fast flight singly or in pairs.
Call: A characteristic .. *chi-chi-chee* .. at flight. A variety of squeaks and chirps while feeding.

Range: Resident with local movements. Western Ghats up to 1800 m (Nilgiris).

Habitat: Evergreen and moist deciduous forests including their stages of degradation, moist scrub and thickets, monocultures (especially those providing nectar such as eucalyptus and arecanut), orchards and in hill / coastal towns.

Family CUCULIDAE

Slender with long graduated tail. Beaks slightly decurved. Two toes pointing forward and the other two behind. Usually solitary. **Parasitic Cuckoos**: Arboreal. Pleasant calls. Often very noisy. Most have mottled, barred and streaked plumage. **Non-parasitic Cuckoos**: Ground birds not commonly ascending trees. Walk and forage among bushes. Flight weak. Low calls or none at all. **Malkohas** have bare faces.

NOTE: Sibley and Monroe (1990) treat Coucals as a separate family, viz. CENTROPODIDAE.

569. Chestnutwinged Cuckoo *Clamator coromandus*
(Redwinged Crested Cuckoo)

Identification: 47 cm. A slender long-tailed crested cuckoo with black body and tail and chestnut-red wings. In flight, crested head, white collar on nape, red wings and whitish underparts diagnostic. **Immature**: Brown above; the feathers broadly edged with rufous. White below.
Call: A harsh rasping cry rendered as .. *creech-creech-creech* ...
Range: Winter visitor. Widespread in Southwestern India known from Maharashtra, Karnataka, Tamil Nadu and Kerala. Regularly on passage in the extreme south.
Habitat: Evergreen and deciduous forests including their stages of degradation and well-wooded urban gardens.

570, 571. Pied Cuckoo *Oxylophus jacobinus* (Pied Crested Cuckoo)

Identification: 33 cm. A slender long-tailed, crested black and white cuckoo of medium size. White patch on wings and white tips to graduated tail diagnostic in flight. **Immature**: Duller black and white with less developed crest. White on wing less extensive. The two races are different only in size. Race *serratus* is larger. Wing over 144 mm. Race *jacobinus* is smaller with wing under 144 mm.
Call: A musical, loud .. *piu* .. *piu* .. *pee-pee-piu* .. *pee-pee-piu* ...

Range: Resident with considerable local movements. Entire Southwestern India up to 2000 m. Race *serratus* visits the northern parts of the Western Ghats (Maharashtra) during the south-west monsoon rains.

Habitat: Deciduous and open secondary forests, scrub and cultivation and also urban gardens.

572. Large Hawk-Cuckoo *Cuculus sparverioides*

Identification: 38 cm. A medium-sized hawk-like cuckoo with short rounded wings. Rather similar to the shikra in overall appearance and flight. Larger size and call separate from common hawk-cuckoo. **Adult**: Ashy-grey and brownish above. Wings and tail brown; the latter banded with black and tipped white. White throat streaked with rufous or grey. Breast rufous. Rest of underparts whitish with brown cross-bands. **Immature**: Dark brown upperparts banded with rufous. Dusky underparts streaked with brown. Whitish tips to tail.

Call: Loud and persistent whistle rendered as .. *pipeeah* .. *pipeeah*..or .. *pipee* .. *pipee*...

Range: Winter visitor on the Western Ghats. Scarce south of Karnataka though reported from Kerala and Tamil Nadu.

Habitat: Forests.

573. Common Hawk-Cuckoo *Cuculus varius*

Identification: 34 cm. A medium-sized hawk-like cuckoo preferring more open habitats than large hawk-cuckoo. Ashy-grey above with white-tipped graduated tail that is relatively shorter and banded pale and blackish. Streaking on throat and breast less distinct or sometimes absent. Bands on underside less distinct. **Immature**: Browner above barred with dull rufous. Barring on tail rufous and black. Buff underparts streaked with brown.

Call: Characteristic .. 'brain fever' .. rendered as .. *wee-piwhit* .. hence the name 'brainfever bird'. Calls at night as well.

Range: Resident with considerable local movements. Entire Southwestern India up to 1000 m.

Habitat: Secondary and deciduous forests, edges of denser evergreen forests, cultivation and scrub, well-planted urban gardens and groves such as tamarind or cashew.

576. Indian Cuckoo *Cuculus micropterus*

Identification: 33 cm. A slender cuckoo separated from the hawk-cuckoos by longer pointed wings and the throat and upper breast being barred (not streaked) or entirely unmarked. Tail with black terminal band. **Adult Male**: Slaty-grey with a brownish tinge above. Underparts white cross-barred with widely separated black bands except on throat and upper breast. Broad black subterminal band on tail diagnostic. **Female**: Rufous-brown tinge on throat and breast. **Immature**: Rufous and white bars on brownish head, nape and upper back. Pale buff below barred with black. Tail rufous with more barring than adult.

Call: Call unmistakable and 4-syllabled. Rendered variously as .. *orange-pekoe* .. or .. *crossword-puzzle* .. or .. *kyphal-pakka* ...

Range: Resident with considerable local movements. Entire Southwestern India.

Habitat: Deciduous and open forests including coastal casuarina groves (Karwar, Karnataka).

578, 579. Common Cuckoo *Cuculus canorus*

Identification: 33 cm. Rather similar to Indian cuckoo except the tail which is darker without a distinct black subterminal band. Tips of tail white. **Female** occurs in two colour phases and **Immature** invariably has a white spot on nape. In hand, grey bars on white wing-edge just after the bend (seen by parting the outer greater wing-coverts) vs full grey in lesser cuckoo is diagnostic. **Adult Female** (hepatic phase): Chestnut-rufous with black bars entire plumage (*See* Lesser Cuckoo, 581, Greybellied Cuckoo, 584 and Banded bay Cuckoo, 582).

Call: Characteristic .. *cuck-koo* ...

Range: Winter visitor. Recorded from western Karnataka (Uttara Kannada and Dakshina Kannada) and Kerala (Trivandrum). Race *bakeri* which is darker than race *canorus,* has been collected in Maharashtra.

Habitat: Secondary evergreen and open forests.

581. Lesser Cuckoo *Cuculus poliocephalus* (Small Cuckoo)

Identification: 26 cm. Small. Not readily distinguished from common cuckoo except by size and darker rump not contrasting against dark tail. **Female**: has a hepatic phase (*See* Common Cuckoo, 578, 579, Greybellied Cuckoo, 584 and Banded Bay Cuckoo, 582). **Immature**: may have a white spot on nape. Call ·most diagnostic.

Call: Rendered as .. *That's your choky pepper* .. *choky pepper* .. or .. *pot-pot* .. *chip-chip-to-you* ...

Range: Winter visitor over the Western Ghats. Known from Gujarat, Maharashtra, northern Karnataka, Kerala and southern Tamil Nadu (Nagercoil).

Habitat: Open forests and urban groves.

582. Banded Bay Cuckoo *Cacomantis sonneratii* (Bay Banded Cuckoo)

Identification: 24 cm. Small. Bright rufous or bay above, including tail, distinctly cross-barred with black. Whitish face and undersides cross-barred with brown. Whitish eyebrow separates from other female cuckoos in hepatic phase. Whitish throat separates from similar hepatic female greybellied cuckoo. **Immature**: More banding on plumage. Often in the company of common ioras (which the species parasitizes) calling characteristically in a harsh tone.

Call: Loud musical .. *wee-wee-wee* .. *weetit* ... *weetit* ... rather persistent.

Range: Resident with local movements. Western Ghats.

Habitat: Edges of evergreen forests, deciduous and open moist forests with scrub and thickets, monocultures and orchards.

584. Greybellied Cuckoo *Cacomantis passerinus* (Indian Plaintive Cuckoo)

Identification: 23 cm. A small active cuckoo. Dark grey-brown fading into whitish towards under tail-coverts. Occasionally full grey. White tips to tail and pale patch on underwing diagnostic in flight. **Female**: in two colour phases. Habits more like a drongo or flycatcher. **Female** (hepatic phase) and **Immature**: Similar to banded bay cuckoo except in having a less pronounced eyebrow and rufous-brown (not whitish) throat.

Call: Mournful high-pitched .. *piteer* .. or .. *kiveer* .. not unlike that of crested goshawk from a distance. Others rendered as .. *weeti-teeti* ...

Range: Resident with considerable local movements. Entire Southwestern India up to 1800 m.

Habitat: Deciduous and open secondary forests with scrub and thickets, cultivation and urban groves.

586. Asian Emerald Cuckoo *Chrysococcyx maculatus*

Identification: 18 cm. A small, brilliant glossy green cuckoo. White underparts barred with green. Chin, throat and breast green as back. Tail with white tips. White patch on primaries conspicuous in flight. **Female**: Green back. Rufous cap and tail; the latter barred with black and tipped white. White underparts (rufous tinged on throat and flanks) barred with brown closely on throat and wider on abdomen.

Call: Described as shrill twitters reminiscent of the vernal hanging parrot.

Range: Winter straggler. Once recorded from Palni hills.

Habitat: Evergreen and secondary forests.

588. Drongo Cuckoo *Surniculus lugubris*

Identification: 25 cm. A small black cuckoo deceptively like a drongo. White under tail-coverts and bases of outer tail feathers (barred) diagnostic. At

rest, characteristic shape of head due to the abrupt union of longish beak and forehead (as in all cuckoos), almost squared tail, less upright sitting posture and in flight, tail and head held above the level of back, especially before landing, separate from drongos of comparable size. **Immature**: Less glossy black with white spots on head, back and breast, besides that on tail.
Call: Rather musical and loud .. *pip-pip-pip-pip-pip-pip* .. repeated often.
Range: Resident with local movements. Entire Southwestern India up to 1500 m.
Habitat: Open forests in the evergreen and deciduous vegetation zones and also urban groves.

590. Asian Koel *Eudynamys scolopacea*

Identification: 43 cm. A familiar medium-sized, slender, full black cuckoo around gardens. Noisy.
Male: Glossy black plumage. Greenish beak. Crimson-red eyes. **Female**: Grey-brown, whitish on belly. Mottled and barred entirely with white and brown. **Immature**: Blackish beak and eyes.
Call: Familiar .. *kuoo* .. *kuoo* ... Others include a rather shrill and piercing .. *kip-kip-kip-kip* .. and .. *uk-keoo-keoo-keoo-keoo-keoo* ...
Range: Resident with local movements. Entire Southwestern India up to 1000 m.
Habitat: Open forests with thicket and scrub, monocultures, orchards and groves, cultivation, mangroves and estuarine coconut plantations, beaches and urban / suburban gardens.

595. Bluefaced Malkoha *Phaenicophaeus viridirostris*
    (Small Greenbilled Malkoha)

Identification: 39 cm. A clumsy ground cuckoo, skulking within bush. Greenish and ashy-grey plumage with long graduated tail broadly tipped white diagnostic. Bright green beak and bluish bare patch around eyes are further pointers. Sexes alike.
Call: Generally very silent. A low call rendered as .. *kraa* ..
Range: Resident. Entire Southwestren India up to 1000 m.
Habitat: Dry and moist scrub and thickets, thorn and *Euphorbia* thickets bordering cultivation and in the outskirts of smaller towns.

598. Sirkeer Malkoha *Phaenicophaeus leschenaultii* (Sirkeer Cuckoo)

Identification: 42–44 cm. A heavy-tailed earthy-brown and rufous ground cuckoo with shiny black streaks on head and breast and bright yellow and red stout beak. In flight white tips to blackish graduated tail diagnostic. Sexes alike.

Call: Normally silent. Call rendered as a sharp .. *kek-kek-kek-kerek-kerek-kerek* .. or .. *kik-kik-kik* ...

Range: Resident. Widespread in Southwestern India up to 1000 m.

Habitat: Dry thorny scrub and thickets.

599. Redfaced Malkoha *Phaenicophaeus pyrrhocephalus*

Identification: 46 cm. A large ground cuckoo within forests with long and broad graduated tail. Glossy green-black with bare crimson-red face and heavy pale green beak diagnostic. White flecks on head. White on tips of tail and abdomen. Sexes alike. **Immature**: More white streaking on black plumage. Shorter tail. Bare patch on face reduced and brick-red.

Call: Usually silent. Calls rendered as .. *kok* .. and .. *kra* ...

Range: Resident. Rare. Localized in southern Kerala and Tamil Nadu (hills of Madurai).

Habitat: Tall evergreen forest.

602. Greater Coucal *Centropus sinensis* (Crow Pheasant)

Identification: 48 cm. An unmistakable glossy black ground cuckoo with chestnut-red wings. Broad graduated tail and weak flight diagnostic. Sexes alike. **Immature**: Do not have the mottled rufous and black plumage as in the northern races.

Call: Characteristic .. *uk-ook-ook* ... Harsh hissing sounds .. *keeaaaah* ...

Range: Resident. Entire Southwestern India up to 2200 m.

Habitat: Evergreen forests (edges), deciduous forests and open secondary thickets, scrub, cultivation, groves and monocultures, mangroves, beaches and urban gardens.

605. Lesser Coucal *Centropus bengalensis*

Identification: 33–37 cm (female larger). A ground cuckoo rather similar to greater coucal though smaller. Whitish tips to tail feathers diagnostic. Sexes alike. **Immature**: Dark brown and rufous. Mottled and barred with black.

Call: Rendered as .. *whoot* .. *whoot* .. *whoot* .. *whoot* .. *kurook* .. *kurook* .. *kurook* .. *kurook* ...

Range: Resident. Localized in western Karnataka, Tamil Nadu and Kerala up to 900 m. Unconfirmed sight record from northern Karnataka.

Habitat: Tall grass and scrub bordering forest.

Family STRIGIDAE

Nocturnal birds. Dull creamy-brown plumage mottled finely or boldly creating a camouflaging effect. Eyes large and frontally placed. Predators with well-developed talons. Soft flight with generally slow wing beats. Eerie calls; combinations of loud hoots and shrieks. Sexes alike. **Barn and Grass Owls:** have a heart-shaped face, slender build and long legs. Pallid coloration. **True Owls:** Variable in size from small owlets to the large horned owls. Owlets, hawk-owl and wood owls have no ear-tufts. Eagle, Fish and scops owls have upright ear-tufts.

NOTE: Some authorities treat Barn, Grass and Bay Owls as belonging to family TYTONIDAE.

606. Barn Owl *Tyto alba*

Identification: 36 cm. A medium-sized white and yellow-brown owl with heart-shaped face and long feathered legs. The yellowish-brown back is mottled finely with black. Underparts spotted with brown. Silent ghost-like flight at night diagnostic. (*See* Eastern Grass Owl, 608)
Call: A shrill shriek, sounding like creaking gears, in flight.
Range: Resident. Entire Southwestern India up to 1000 m.
Habitat: Urban and suburban buildings and ruins.

608. Eastern Grass Owl *Tyto longimembris*

Identification: 36 cm. A pallid owl very similar to barn owl. Darker brown back with white spots, black spots in front of eyes, longer stockinged legs and white tail cross-barred with buff identify the grass owl. In flight appears whitish.
Call: Described as similar to that of barn owl.
Range: Resident. Known from the Nilgiris, High Range of Kerala, Brahmagiris (Karnataka) and Palni hills up to 1800 m.
Habitat: Tall grass along the folds and slopes.

610a. Oriental Bay Owl *Phodilus badius*

Identification: 29 cm. A small chestnut-bay owl with short ears. Chestnut above spotted with white and black. Chestnut tail barred with black. Face pinkish with a white and black ruff bordering. Dark rufous marks around eyes diagnostic. Pinkish below spotted with black and white on breast and flanks.
Range: Nelliampathy Hills and Anaimalais in Kerala.
Habitat: Evergreen forests and secondary moist forests.

614. Pallid Scops Owl *Otus brucei*

Identification: 22 cm. A small, slim greyish owl with indistinct ear-tufts. Greyish-brown above with a sandy-yellow tinge finely vermiculated and streaked with black. Whitish underparts heavily streaked with black.

Call: Described as similar to that of Eurasian scops owl.
Range: Winter visitor. Mumbai, Pune, Thane and Ratnagiri (Maharashtra).
Habitat: Groves in dry terrain.

615, 617. Eurasian Scops Owl *Otus scops*

Identification: 19 cm. The smallest of our scops owls. Rufous or grey
plumage (the latter commoner). Ear-tufts conspicuous when alarmed. Yellow eyes and distinct markings on underparts identify this species. In rufous
phase whitish eyebrows diagnostic. Races separable only with difficulty
even in hand. Race *pulchellus* has the third primary feather longest (vs.
fourth in race *rufipennis*).
Call: Rendered variously as .. *toink-tonk-ta-tonk* .. *kurook-took* .. *wukh-tuk-tah*
.. and *wuck-chug-chug* ... Confusable with the call of grey nightjar.
Range: Resident. Western Ghats. Race *pulchellus* has been recorded as a
winter vagrant in Mumbai.
Habitat: Open forest (evergreen and deciduous), groves and orchards in the hills.

622, 623. Indian Scops Owl *Otus bakkamoena* (Collared Scops Owl)

Identification: 23–25 cm. A small horned owl in rufous-brown or grey-brown
colour phases. Pale collar on hindneck diagnostic. Underparts paler-whitish.
Entire plumage mottled and with fine barring. Race *bakkamoena* has darker
plumage than race *marathae*.
Call: Single soft .. *wut* ... Another described as a slow series of .. *ack* .. *ack*
.. *ack* ...
Range: Resident. Entire Southwestern India up to 1200 m. The race *bakkamoena* occurs south of Gujarat through Maharashtra down to Kerala.
Habitat: Open forests, groves, cultivation and urban groves.

627. Eurasian Eagle Owl *Bubo bubo* (Great Horned Owl)

Identification: 56 cm. A large brown owl with distinct ear-tufts. Cat-like
appearance from a distance and in silhouette diagnostic. Entire plumage
heavily streaked with black. White throat patch. Yellow-orange eyes. Fully
feathered legs.
Call: Loud and resonant .. *bu-bo*... Also rendered as .. *dur-goon* ... and ..
to-whoot ...
Range: Resident. Entire Southwestern India.
Habitat: Open deciduous forests, cultivation and rocky foothills covered with
grass and scrub.

628. Spotbellied Eagle Owl *Bubo nipalensis* (Forest Eagle Owl)

Identification: 63 cm. The largest of the eagle owls. Heavy build and distinct (slanting) black and white ears-tufts. Fully feathered legs and brown eyes diagnostic. **Adult**: Brown. Back mottled with buff. White below barred only on throat and breast. Blackish spots or drops on rest of underparts. **Immature**: Buffy-white, paler on head. Barred with brown on entire plumage.

Call: Described as a loud .. *boom* ... Also called the 'devil bird' for the so-called diabolical blood-curdling shrieks, resembling those of a woman being strangled, that the owl is said to produce.

Range: Resident. Western Ghats southwards from Belgaum (Karnataka) up to 2100 m.

Habitat: Dense evergreen and moist deciduous forests.

630. Dusky Eagle Owl *Bubo coromandus* (Dusky Horned Owl)

Identification: 58 cm. A large horned owl similar to the Eurasian eagle owl. Erect ear-tufts standing close to each other, pale yellow eyes and grey-dusky plumage are diagnostic.

Call: A loud resonant .. *wo* .. *wo* .. *wo* .. *woo-o-o-o*.. Uttered even during the day.

Range: Resident. Known from Nilgiris and north up to Karwar (Karnataka).

Habitat: Open forests, groves and riverine vegetation.

631. Brown Fish Owl *Ketupa zeylonensis*

Identification: 56 cm. A large brown horned owl with unfeathered legs. A distinct white patch on throat. Rufous-brown above and whitish below. Entire plumage streaked and mottled with black. Golden-yellow eyes.

Call: Described as a hollow sounding .. *boom-boom* .. or .. *boom-o-boom* ...

Range: Resident. Entire Southwestern India up to 1400 m.

Habitat: Open forests, groves and orchards bordering marshes even in the neighbourhood of human habitation.

636, 637. Jungle Owlet *Glaucidium radiatum* (Barred Jungle Owlet)

Identification: 20 cm. A small, short-tailed, round-headed owl without ear-tufts. Dark brown above barred with rufous. White moustachial streak. Underparts largely white barred with olive-brown on flanks and under tail-coverts. Rufous patch on underwing diagnostic in flight. The race *malabaricum* has a richer rufous plumage than race *radiatum*.

Call: Characteristic loud .. *kao* .. *kao* .. *kao* .. *kaokuk* .. *kao* .. *kuk, kao- kuk* ... *kao* .. *kuk* rising and gradually fading away. Another call rendered as a pleasant .. *woioioioioioi* .. *keek* ...

Range: Resident. Western Ghats up to 2000 m. The race *radiatum* displaces race *malabaricum* north of Goa and in the east including the Nilgiris.
Habitat: Evergreen and deciduous forests with their stages of degradation.

642, 644. Brown Hawk-Owl *Ninox scutulata*

Identification: 32 cm. A hawk-like medium-sized owl without ear-tufts. Grey-brown above with whitish forehead and spots on shoulders. Brownish streaked throat and foreneck. Rest of underparts white with large reddish-brown drops. Barred tail tipped with white. Race *lugubris* is paler with a longer wing (over 208 mm). Race *hirsuta* is darker on head and underparts and with profuse barring on underwing.
Call: A soft .. *oo* ...*uk* .. *oo* .. *uk* .. continuously uttered.
Range: Resident. Western Ghats up to 1300 m. The northern race *lugubris* is separated from the southern race *hirsuta* at Mumbai.
Habitat: Open forests and edges of tall forests along rivers.

652. Spotted Owlet *Athene brama*

Identification: 21 cm. A familiar small squat round-headed owl without ear-tufts. Grey (spotted white) with white face, round yellow eyes and whitish underparts cross-barred with black, diagnostic.
Call: A series of harsh squeaks and scoldings rendered as .. *chirrur* .. *chirrur* .. *chirrur* .. *cheevak* .. *cheevak* .. *cheevak* ...
Range: Resident. Entire Southwestern India.
Habitat: Open cultivation and scrub, estuaries and beaches, always close to human habitation and frequently within buildings.

657. Mottled Wood Owl *Strix ocellata*

Identification: 48 cm. A largish round-headed owl with no ear-tufts. Largely yellowish-buff finely mottled with black, brown and white. Yellowish-buff wings diagnostic in flight.
Call: An eerie loud .. *hooot* ... Also a laughter-like .. *chuhua-aa* ...
Range: Resident. Widespread in Southwestern India.
Habitat: Open forests and groves bordering cultivation and also within town limits.

659. Brown Wood Owl *Strix leptogrammica*

Identification: 47 cm. A largish chocolate-brown and buffish round-headed owl without ear-tufts. Whitish face. White throat. Entire plumage barred closely with pale and dark brown.
Call: Rendered as a low .. *tok* .. *tu-hoo* ...

Range: Resident. Widespread on the Western Ghats southwards from Mahabaleshwar (Maharashtra) through Uttara Kannada (Karnataka) and Kerala up to 1800 m.

Habitat: Dense evergreen, semi-evergreen and moist deciduous forests.

664. Shorteared Owl *Asio flammeus*

Identification: 38 cm. A medium-sized slim owl with short, dark and erect ear-tufts (visible only at close range). Pale buff plumage streaked with dark brown; darkest on head. Whitish face with darker ruff. Longitudinal brown streaks on breast. In flight rufous and white wings with a dark bar across diagnostic.

Range: Winter visitor. Widespread on the Western Ghats up to 1400 m.

Habitat: Grass and scrub.

Family PODARGIDAE

Small nocturnal birds with cryptic coloration. Grey or brown with fine mottling and vermiculation. Thick beaks with wide frog-like gapes. Arboreal. Sit upright unlike nightjars.

NOTE: Sibley and Monroe (1990) treat Indian Frogmouths as belonging to family BATRACOSTOMATIDAE.

666. Ceylon Frogmouth *Batrachostomus moniliger*

Identification: 23 cm. A cryptically coloured owl-nightjar-like bird with broad thick beak, an exceptionally wide gape and large yellow eyes. Sitting motionless and upright on low branches, characteristic. **Male**: Greyish mottled with white, buff and black often 'in the form of blotches. **Female**: Rufous-brown with white spots and blotches.

Call: Rendered as .. *kooroo* .. *kooroo* .. *kooroo* .. and .. *whoo* .. *whoo* .. *whoo* .. *whoo* ...

Range: Resident. Patchily distributed over the Western Ghats up to 1200 m. Known from Uttara Kannada (Karnataka) and Silent Valley, Periyar, Trichur and Trivandrum districts in Kerala.

Habitat: Evergreen forests including their secondary stages.

Family CAPRIMULGIDAE

Nocturnal birds with soft cryptically coloured plumage. Brown mottled with buff, white and black. Long-tailed with long, slender wings. Small beaks and legs. Most have white-buff patches on wings; conspicuous in flight. Sit horizontally on ground; rarely across branches. Calls most diagnostic.

NOTE: Sibley and Monroe (1990) treat the Great Eared Nightjar as belonging to family EUROSTOPODIDAE.

669. Great Eared Nightjar *Eurostopodus macrotis*

Identification: 41 cm. A large nightjar with ear-tufts. In flight, long tail and lack of pale patches on wings and tail striking.
Call: Loud whistle rendered as .. *whi-wheeeew* ...
Range: Resident. Patchily distributed over the Western Ghats up to 1000 m. Known from Kerala (Kottayam, Quilon and Trivandrum districts, Periyar and Silent Valley) and Goa.
Habitat: Evergreen and moist deciduous forests including their stages of degradation.

671. Grey Nightjar *Caprimulgus indicus* (Jungle Nightjar)

Identification: 29 cm. A medium-sized nightjar. Greyish plumage without a pale hindcollar. White tips on 4 pairs of outer tail feathers and almost fully feathered legs diagnostic. **Female**: Differs from male in that buff replaces white on wings and tail.
Call: A pleasant *uk-krukroo*... Others rendered as .. *chuck* .. *chuck* .. *chuck* .. *chuck* .. and .. *chuckoo* .. *chuckoo* .. *chuckoo* ... More than one of these calls may be heard locally. (*See* Eurasian Scops Owl, 615, 617)
Range: Resident with local movements. Western Ghats up to 2300 m.
Habitat: Deciduous forests, teak plantations, scrub and cultivation.

673. Eurasian Nightjar *Caprimulgus europaeus* (European Nightjar)

Identification: 25 cm. A greyish nightjar, smaller than grey nightjar. White on two outer pairs of tail feathers and first three primaries diagnostic. Legs almost fully feathered. **Female**: Lacks the white on tail. White on wings duller.
Call: Not familiar. Rendered as .. *sweeesh* .. and .. *chuck-chuck-chuck* ...
Range: Winter straggler recorded in Gujarat and Mumbai.
Habitat: Open forests and scrub on hillsides.

674. Sykes' Nightjar *Caprimulgus mahrattensis*

Identification: 23 cm. A small sandy-grey nightjar. Plumage finely vermiculated with black. Unfeathered legs. Outer tail feathers tipped white. Semidesert. **Female**: White on wings duller and on tail vermiculated with brown.
Call: Described as that of a frog. Also rendered as .. *prrrrrr* ...

Range: Winter straggler along the drier margins of the Western Ghats south to about Belgaum.
Habitat: Stony wasteland and open hillsides.

676. Largetailed Nightjar *Caprimulgus macrurus* (Longtailed Nightjar)

Identification: 28 cm. A medium-sized brownish-buff forest nightjar. White throat. White spot on first 4 primaries (vs. 3 in Eurasian nightjar). White tips to 2 outer pairs of tail feathers. Legs fully feathered. Often on forest paths at dusk. Eyes beam red against vehicle headlights.
Call: Rendered as a loud .. *chaunk* .. *chaunk* .. *chaunk* .. However, the calls most frequently heard in the forests of northern Karnataka are identical to those described for the Ceylonese race .. *grog-groggrog* .. and .. *quoffr* .. *quoffr* ...
Range: Resident. Southern Maharashtra through Karnataka over the entire Western Ghats up to 2000 m.
Habitat: Secondary evergreen forests (edges), open deciduous forests, teak plantations, bamboo facies, scrub and cultivation.

680. Indian Nightjar *Caprimulgus asiaticus*

Identification: 24 cm. A small nightjar distinctly streaked with black on back. Unfeathered legs diagnostic. White patches on either side of throat. Separated from large tailed nightjar by shorter tail and stockier appearance. Darker plumage separates from Sykes' nightjar.
Call: Familiar *chuck-chuck-chuck-chuck-r-r-r*...
Range: Resident with local movements. Entire Southwestern India up to 1500 m.
Habitat: Open rocky scrub, lateritic outcrops, cultivation and urban wastelands.

682. Savanna Nightjar *Caprimulgus affinis* (Franklin's Nightjar)

Identification: 25 cm. A small nightjar. Less boldly streaked plumage and a pale 'V' on back diagnostic at rest. Two outer pairs of tail feathers fully white except on tip. White patch on throat parted in the centre. Unfeathered tarsus. **Female**: More rufous than male lacking the white tail feathers. Separated from female longtailed nightjar by uniformly marked tail without distinct paler tips.
Call: A sharp whiplash .. *sweeesh* .. Also rendered as .. *cho-ee* ...
Range: Resident with local movements. Western Ghats up to 1800 m.
Habitat: Edges of secondary forests, open hillside scrub and cultivation.

Family APODIDAE

Slender birds. Long wings arched and bow-like in flight. Weak legs and short beaks. Fast flying. Feed aerially. Gregarious. **Swifts and Swiftlets**: short-tailed with or without forks. Colonial nests of mud, feathers and saliva. Clinging posture at rest. **Tree Swifts**: slender, long tail with a deep swallow-like fork. Erect crest from base of beak visible at rest. Perches upright. Shallow nests of lichen and fibre on branches.

NOTE: Some authorities treat Crested Tree Swift as belonging to family HEMIPROCNIDAE.

685. Indian Swiftlet *Collocalia unicolor* (Edible-nest Swiftlet)

Identification: 12 cm. A tiny sooty-brown swift with slightly forked tail. Rump not paler than back or abdomen diagnostic in flight.
Range: Resident. Western Ghats south of Ratnagiri (Maharashtra) up to 2200 m.
Habitat: Over evergreen forests, exposed hilltops, streams, open grassland and in the vicinity of rocky caves.

691. Brownbacked Needletail *Hirundapus gigantea*
(Brownthroated Spinetail Swift)

Identification: 23 cm. The largest of our swifts. Blackish-brown with short square tail and needles projecting behind. White under tail-coverts contrasting with dark body. Fast flight often with a 'whizz' at close range diagnostic.
Range: Resident. Western Ghats. Goa to the southernmost hills of Kerala up to over 1000 m.
Habitat: Evergreen and secondary moist forests, moist thickets and hillside grass.

692. Whiterumped Spinetail *Zoonavena sylvatica*

Identification: 11 cm. A tiny forest swift. Sooty-black with white rump, abdomen and under tail-coverts. White on underparts separate from otherwise similar little swift.
Range: Resident. Western Ghats. Goa to southern Kerala up to over 1000 m.
Habitat: Evergreen and moist deciduous forests including their secondary stages and associated scrub.

694. Alpine Swift *Tachymarptis melba*

Identification: 22 cm. A large, brownish long-winged swift with slightly forked tail. Dark breast-band and under tail-coverts contrasting against the white underparts, diagnostic in flight.
Range: Resident. Entire Southwestern India up to over 1000 m.
Habitat: Tall riverine cliffs, open forests and marshes.

700. Forktailed Swift *Apus pacificus* (Large Whiterumped Swift)

Identification: 15 cm. A black swift with white rump, rather similar to house swift. Forked tail diagnostic.
Range: Winter visitor. Mumbai, Ratnagiri (Maharashtra), Uttara Kannada, Dakshina Kannada (Karnataka) and Malabar (Kerala).
Habitat: Coastal towns.

703, 704. Little Swift *Apus affinis* (House Swift)

Identification: 15 cm. A familiar black swift with white rump. Chin and throat white. Squared tail diagnostic. The birds in Kerala have been considered as belonging to the Sri Lankan race *singalensis* due to the darker chin and throat than in race *affinis*.
Call: Lively chirpings as the birds wheel around a roost.
Range: Resident. Entire Southwestern India up to 2000 m.
Habitat: Urban, beaches, estuaries, cultivation, dams and under bridges in the higher hills (Nilgiris).

707. Asian Palm Swift *Cypsiurus batasiensis*

Identification: 13 cm. A slender, small, sooty-brown swift with long deeply forked tail. In overhead flight, tail is closed giving the bird a bow–arrow appearance. In the vicinity of palms.
Range: Resident. Entire Southwestern India up to 1000 m.
Habitat: Open forests in close association with palms such as *Borassus, Corypha* and the cultivated arecanut, over marshes and near human habitation.

709. Crested Tree Swift *Hemiprocne coronata*

Identification: 23 cm. A slender blue-grey long-tailed swift appearing more like a swallow in overhead flight. Sits upright on overhead power or telegraph lines. Long tail and erect crest diagnostic at rest. Piercing call reminiscent of shikra diagnostic. **Male**: Chestnut sides of face, chin and throat.
Call: Characteristic .. *pick-wick* .. *pick-wick* .. audible over a distance.
Range: Resident with local movements. Western Ghats up to 1200 m.
Habitat: Open secondary and degraded forests and associated thickets, monocultures, etc., often close to human habitation.

Family TROGONIDAE

Brightly coloured forest birds with short rounded wings and long graduated square-cut tails. Short beaks. Bluish bare patch around eyes. Sexes different.

710, 711. Malabar Trogon *Harpactes fasciatus*

Identification: 31 cm. A bright forest bird with black head, rufous back and brilliant crimson-red underparts. Chestnut-red tail with black border. Outer tail feathers white and conspicuous when spread. Fine black vermiculation on silvery-grey wings. **Female**: Duller. Brownish head. Rufous-orange underparts. The race *legerli* has more extensive white on wings, than race *malabaricus*.

Call: Soft .. *cue* .. or .. *tue* ... *krr-r-r-r* .. when alarmed.

Range: Resident. Western Ghats up to 1500 m. The southern race *malabaricus* occurs throughout the range except in Surat Dangs where it is replaced by race *legerli*.

Habitat: Evergreen and moist deciduous forests including their stages of degradation, teak plantations and other neglected tree plantations in the hills.

Family ALCEDINIDAE

Large-headed brightly coloured birds with long, dagger-like beaks. Short legs and tail. Found in the vicinity of water. Plunge into water for fish. Solitary. Sexes alike.

NOTE: Sibley and Monroe (1990) treat the genera *Halcyon* and *Pelargopsis* as family DACELONIDAE and *Ceryle* as CERYLIDAE.

719, 720. Pied Kingfisher *Ceryle rudis*

Identification: 31 cm. A medium-sized, speckled and barred, black and white kingfisher. Small nuchal crest. Distinct black bands across white breast. Characteristic hovering over water. Pairs. **Female**: Differs from male in having a single breast-band. The races differ in the intensity of black on upperparts, race *travancorensis* having darker upperparts spotted with white and more extensive black spotting on flanks, than race *leucomelanura*.

Call: A sharp .. *chirruk* .. *chirruk* .. uttered on wings.

Range: Resident. Entire Southwestern India up to 1500 m. The race *travancorensis* occurs southwards from Calicut in Kerala and Tamil Nadu.

Habitat: Marshes including paddyfields, rivers, streams, lakes, estuaries, mangroves, salt pans, beaches and ditches within urban limits.

723, 724. Common Kingfisher *Alcedo atthis* (Small Blue Kingfisher)

Identification: 18 cm. A small greenish-blue and orange kingfisher perched low beside water. Often bobs its head. In flight brilliant blue back and characteristic call diagnostic. Solitary. The races differ chiefly in intensity of colour, race *taprobona* being darker and more bluish than green as against race *bengalensis*.

Call: Shrill .. *chichee* .. *chichee* .. or .. *chichichee* ...

Range: Resident. Entire Southwestern India up to 1800 m. The race *taprobana* is more widespread on the Ghats intergrading with *bengalensis* at Gujarat.

Habitat: Marshes including rain puddles, ditches and paddyfields, channels, rivers, streams, estuaries, mangroves, salt pans, beaches and also within urban limits.

725, 726. Blue-eared Kingfisher *Alcedo meninting*

Identification: 16 cm. A small kingfisher confusable with small blue kingfisher. Darker purplish-blue upperparts, blue ear-coverts (*vs.* orange-brown) and preference for forests diagnostic. (*See* Common Kingfisher, 723, 724)

Call: Similar to that of common kingfisher. Described as somewhat sharper.

Range: Resident. Western Ghats. Belgaum southwards till the southern tip up to 1500 m. The smaller and darker Sri Lankan race *phillipsi* presumably occurs in the Western Ghats as well as race *coltarti.*

Habitat: Hill-streams in evergreen forests and also well-watered mixed orchards of arecanut and cacao.

727. Blackbacked Kingfisher *Ceyx erithacus* (Threetoed Kingfisher)

Identification: 13 cm. A tiny, brilliantly coloured forest kingfisher with bright coral-red beak, purple-blue-black or violet back, pinkish rump and orange underparts. Solitary in dense vegetation.

Call: Described as similar to that of common kingfisher .. *chichee* .. *chichichee* ...

Range: Resident with local movements during the rains. Western Ghats up to 1000 m.

Habitat: Hill-streams in dense moist forests, evergreen forests (away from water) and well-watered mixed orchards of arecanut and cacao.

730. Storkbilled Kingfisher *Pelargopsis capensis*

Identification: 38 cm. A large blue kingfisher with a massive red beak. Brown head and brown-yellow underparts diagnostic. (*See* Whitethroated Kingfisher, 736)

Call: A variety of harsh chuckles .. *ke-ke-ke-ke* .. not unlike that of the Indian roller and tree pie. A pleasant .. *peer-peer-pur* .. rather similar to that of the Indian cuckoo from a distance.

Range: Resident. Western Ghats up to at least 1000 m.

Habitat: Rivers and streams through evergreen and deciduous forests, reservoirs lined with tall vegetation, coastal paddyfields and other wetlands close to hill settlements, estuaries and mangroves.

736. Whitethroated Kingfisher *Halcyon smyrnensis*
(Whitebreasted Kingfisher)

Identification: 28 cm. A medium-sized brilliant blue chocolate-brown and white kingfisher. Brown head, white breast and long red beak diagnostic. White patches on wings distinct in flight. Away from water.

Call: A harsh and piercing .. *kek-kek-kekik-kekik* .. *ke-ke-ke-ke* .. at flight. A rather solemn .. *kililili* .. at rest.

Range: Resident with local movements. Entire Southwestern India up to 2000 m.

Habitat: Wetlands of all kinds including rain puddles, estuarine marshes, etc. This species is not tied to wet environs as much as the other kingfishers, and hence is commonly seen in dry cultivation, open forests, meadows, roadsides, sandy beaches and urban gardens.

739. Blackcapped Kingfisher *Halcyon pileata*

Identification: 30 cm. A medium-sized deep cobalt-blue kingfisher with black cap and white-rusty-brown underparts. Bright red beak and white collar between black cap and blue upperparts. White patch on blackish wings diagnostic in flight. Estuarine. (*See* Whitethroated Kingfisher, 736)

Call: Rather similar to that of whitethroated kingfisher but shriller.

Range: Resident with local movements. Mumbai southwards along the coast, occasionally straying inland (Bangalore).

Habitat: Beaches, estuaries, mangroves and hill-streams within evergreen forest environs.

740. Collared Kingfisher *Todirhamphus chloris* (Whitecollared Kingfisher)

Identification: 24 cm. A small kingfisher, intermediate in size between common kingfisher and whitethroated kingfisher. Greenish-blue crown and upperparts. White eyebrow, collar and underparts (streaked in immature). Black beak and line through eyes diagnostic.

Call: Rendered as a harsh .. *krerk-krerk-krerk* ...

Range: Resident (?). Rare and local. Ratnagiri district (coastal Maharashtra) and Kerala.

Habitat: Mangrove swamps.

Family MEROPIDAE

Active green insect-eating birds. Long curved beaks. Elongated central tail feathers typical for most species. Pointed reddish wings and characteristic calls diagnostic in overhead flight. Sexes alike.

744. Chestnutheaded Bee-eater *Merops leschenaulti*

Identification: 21 cm. A small grass-green bee-eater with chestnut-red head and upper back. Yellow chin and throat bordered by a black necklace. Lack of elongate tail feathers diagnostic in overhead flight. Flocks.
Range: Resident with local movements. Southwards from Ratnagiri district (southern Maharashtra) over the Western Ghats up to 1500 m.
Habitat: Edges of evergreen forests, open secondary forests with scrub, forest monocultures, deciduous forests, along streams and in coastal vegetation.

747. Bluecheeked Bee-eater *Merops superciliosus*

Identification: 31 cm. A medium-sized bee-eater with long central tail feathers. Readily distinguished from the little green bee-eater by larger size and chestnut throat and from the bluetailed bee-eater by green rump and tail. Flight more sailing than the little green bee-eater's. (*See* Bluetailed Bee-eater, 748)
Range: Straggler during winter migration through Gujarat and Maharashtra as far south as Mumbai.
Habitat: Marshy habitats.

748. Bluetailed Bee-eater *Merops philippinus*

Identification: 31 cm. A medium-sized bee-eater with elongated tail feathers. Blue rump and tail diagnostic. Flocks.
Range: Winter visitor. Entire Southwestern India up to 1000 m.
Habitat: Open forest, marshes, cultivation, and within urban limits.

750. Little Green Bee-eater *Merops orientalis*

Identification: 21 cm. A small bee-eater with elongated tail feathers. Golden-rufous crown and bluish-green throat with a black border diagnostic even when individuals lack the long tail feathers. Flocks on fences and overhead wires.
Call: A shrill .. *tree-tree-tree* .. *tit-tit-tit* ...
Range: Resident with local movements. Entire Southwestern India up to 2000 m.
Habitat: Open forests, plantations, orchards, cultivation, scrub, beaches, estuaries, all kinds of wetlands and urban roadsides and wastelands.

753. Bluebearded Bee-eater *Nyctyornis athertoni*

Identification: 36 cm. A large green forest bee-eater without elongated tail feathers. Bluish crown, bright blue (streaked) throat and breast and yellowish underparts streaked green diagnostic. Flight like that of brownheaded barbet or Malabar grey hornbill; head dipping. Solitary or pairs.
Call: A raucous .. *kor-r-r* .. *kor-r-r* .. reminiscent of the storkbilled kingfisher from a distance.
Range: Resident. Western Ghats up to 1700 m (Nilgiris).
Habitat: Secondary open evergreen and moist deciduous forests.

Family CORACIIDAE

Big-sized birds with thick heads and stout beaks. Bright blue rounded wings, slow flapping flight and impressive acrobatic flight displays. Perch on overhead telephone wires. Pairs or solitary. Noisy. Sexes alike.

754. European Roller *Coracias garrulus* (Kashmir Roller)

Identification: 31 cm. A largish bird with blue head and underparts. Brownish back. Uniformly blue-black flight feathers and overall paler coloration separate from the commoner European roller.
Range: Winter visitor. Coastal Maharashtra southwards to Bhatkal (coastal Karnataka).
Habitat: Open meadows and cultivation.

756. Indian Roller *Coracias benghalensis*

Identification: 31 cm. A bright blue and brown large-headed bird perched on overhead wires and bare trees. Purplish streaked throat and breast (at rest) and bright blue wings alternately banded with darker blue on flight feathers (in flight) separate from European roller.
Call: A series of harsh raucous sounds .. *keak-keak-ke-ke-ke* .. *kek-kek* .. and .. *kek-ke-ke-ke-ke* .. *kek-ke-ke-ke-ke* .. at rest and in flight.
Range: Resident with local movements. Southwestern India up to 1000 m.
Habitat: Open forests, cultivation, marshes, estuaries, salt pans, beaches and urban wastelands.

759. Dollarbird *Eurystomus orientalis* (Broadbilled Roller)

Identification: 31 cm. A blackish-blue and purple roller with black head and thick coral-red beak. A pale round bluish patch on dark wings diagnostic in flight. Solitary or pairs on tall exposed perches. Crepuscular.
Call: Rendered as a raucous .. *chaak-chaak* .. *chaak* ...

Range: Resident. Local. Western Ghats. Coorg (Karnataka) southwards through Nilgiris and Kerala up to 500 m.
Habitat: Clearings and cultivation in evergreen forest environs.

Family UPUPIDAE

Bright rufous, black and white birds. Long slender decurved beaks. Fan-like crest. Ground feeding. Sexes alike.

764, 765. Eurasian Hoopoe *Upupa epops*

Identification: 31 cm. An elegant slender, rufous, black and white ground bird with long slender beak and fan-like crest. Bright black and white wings and tail diagnostic in flight. The race *ceylonensis is* darker with no white spots on crest. as against race *saturata*.
Call: Characteristic .. *oop-oop-oop* ... Pairs utter a wheezing .. *see* .. *see* .. *see* ...
Range: Resident with local movements. Entire Southwestern India up to over 2000 m. In winter, the migrant race *saturata* mixes with race *ceylonensis* in Gujarat and Maharashtra.
Habitat: Open forests, cultivation, estates and hill gardens, beaches, urban gardens and buildings.

Family BUCEROTIDAE

Large grey-black and white birds. Massive beaks with horns or casques diagnostic. Characteristic flight on broad rounded wings. Long-tailed. Hole-nesting. Females sealed within till young are partially grown. Sexes not strikingly different.

767. Indian Grey Hornbill *Ocyceros birostris*

Identification: 61 cm. A brownish-grey kite-sized bird with heavy, curved blackish beak. A small black casque and white tips to blackish tail are diagnostic. **Female**: Smaller casque. **Immature**: No casque. Yellow beak confusable with Malabar grey hornbill. However lack of white tips to primaries is diagnostic.
Call: Normally a squeal like that of kites .. *wheeee* ... Others rendered as .. *k-k-k-ka-e* ...
Range: Resident with local movements. Entire Southwestern India along the margins up to 1000 m including Kerala (Trichur–Palghat) and coastal Maharashtra (Raigad).

Habitat: Open moist forests with scrub, dry deciduous forests, scrub and cultivation and also urban groves.

768. Malabar Grey Hornbill *Ocyceros griseus*

Identification: 59 cm. A small slaty-grey hornbill without a casque. Yellow beak, streaked head, neck and breast and white tips to black primaries diagnostic.
Call: A loud cackle .. *kyah-kyah-kyah* .. *kyah* .. reminiscent of a village hen being caught.
Range: Resident. Widespread on the Western Ghats south of Mumbai up to 1600 m.
Habitat: Evergreen and moist deciduous forests including their stages of degradation and associated monocultures.

775. Malabar Pied Hornbill *Anthracoceros coronatus*

Identification: 92 cm. A large black and white hornbill with a distinct black and yellow casque. Yellow beak. Fully white outer tail feathers diagnostic. Smaller size, faster wing-beats and predominantly black plumage separate from great hornbill in flight.
Call: Loud yelps similar to that of a hurt puppy.
Range: Resident with local movements. Ratnagiri (Maharashtra) southwards over the Western Ghats up to about 600 m.
Habitat: Open forests, monocultures and thickets interspersed with tall trees, estuarine vegetation, cultivation lined with tall trees and suburban groves.

776. Great Hornbill *Buceros bicornis*

Identification: 130 cm. A huge white and black hornbill with massive yellow beak and a concave surfaced casque. Face, wings and underparts black. In flight, white neck, black wings with two white bands and white tail with black band diagnostic. Large size and slow wing-beats often with a distinct 'whizz' are further pointers when overhead.
Call: Loud raucous .. *aang* .. similar to that of a large goose. A single low .. *tok* .. often repeated as if a hollow log is being hammered.
Range: Resident with local movements. Western Ghats. Khandala (Maharashtra) southwards till the southern tip up to 1500 m.
Habitat: Tall evergreen and moist deciduous forests.

Family CAPITONIDAE

Small to medium-sized green birds with large stout and conical beaks. Long bristles from base of beak. Two toes point backwards as in woodpeckers. Largely frugivorous. Incessant croaking calls. Hole-nesting. Sexes alike.

NOTE: Sibley and Monroe (1990) treat Barbets as family MEGALAIMIDAE.

781, 782. Brownheaded Barbet *Megalaima zeylanica* (Green Barbet)

Identification: 27 cm. The largest of our barbets. Green. Brownish, streaked head, neck and breast. Massive yellow beak and bare yellow skin around eyes diagnostic. Undulating flight, as if the head is pulling the bird down, is a further pointer. The race *zeylanica* is darker in coloration than race *inornata*.
Call: Loud .. *kor-r-r-r-r* .. *kotrook* .. *kotrook* ... Louder and more resonant than whitecheeked barbet. Also a softer low *oo-oo-ook*...
Range: Resident. Entire Southwestern India up to 1500 m. Race *zeylanica* occurs south of the Palghat gap.
Habitat: Open and deciduous forests, plantations, orchards and urban groves.

785. Whitecheeked Barbet *Megalaima viridis* (Small Green Barbet)

Identification: 23 cm. A medium-sized green barbet rather similar to the brownheaded barbet. Smaller beak, and white patches above and below eyes are diagnostic.
Call: Loud .. *kor-r-r-r* .. *kotroo* .. *kotroo* .. sharper and more blunt than that of brownheaded barbet. Another call often heard is a soft .. *keea-a-ah* .. *keea-a-ah* ...
Range: Resident. Entire Southwestern India up to over 2000 m.
Habitat: Evergreen and moist deciduous forests including their associated secondary stages, monocultures, orchards, estuarine vegetation and urban gardens.

790. Crimsonfronted Barbet *Megalaima rubricapilla*
(Crimsonthroated Barbet)

Identification: 17 cm. A small, dumpy, green forest barbet. Short tail. Crimson forehead, face, throat and breast are diagnostic. (*See* Coppersmith Barbet, 792)
Call: An abrupt .. *wut* .. or .. *tuk* .. repeated at a higher frequency than that of coppersmith barbet.
Range: Resident. Western Ghats. Southern Maharashtra–Goa southwards through Nilgiris and Palnis till the southern tip up to 1200 m.
Habitat: Evergreen forests and associated vegetation.

792. Coppersmith Barbet *Megalaima haemacephala*
(Crimsonbreasted Barbet)

Identification: 17 cm. A small familiar countryside barbet often referred to as 'coppersmith' due to its incessant calling. Dumpy and green. Yellow throat and face bordered with black diagnostic. Forehead and breast crimson.
Call: Characteristic incessant .. *tuk* .. *tuk* ...
Range: Resident. Entire Southwestern India up to 1000 m.
Habitat: Open deciduous forests, orchards, village and urban groves.

Family PICIDAE

Small to large and often brightly coloured birds. Two toes pointing backwards. Forage on trunks and branches of trees; rarely on ground. **Wrynecks**: camouflaging nightjar-like colour pattern. **Piculets**: tiny and short-tailed moving about branches like a nuthatch. **Woodpeckers**: Bright colouration with prominent crests. Loud calls and drumming characteristic. All are hole-nesting.

796. Eurasian Wryneck *Jynx torquilla*

Identification: 19 cm. A slim silvery-grey and brown bird streaked and vermiculated like a nightjar. Feeds among low bushes often on ground, hopping about with tail cocked. Flight similar to sparrow. Also confusable with clamorous warbler when moving within and flying out of a bush. Dry habitats.
Call: Rendered as a .. *chewn* .. *chewn* .. *chewn* .. similar to that of common myna and blackrumped flameback.
Range: Winter visitor. Widespread in Southwestern India.
Habitat: Thorn and scrub jungle and cultivation.

799. Speckled Piculet *Picumnus innominatus*

Identification: 10 cm. A tiny, yellowish-olive woodpecker. White underparts spotted and barred with black. Short black and white tail. Dark blackish band through eyes bordered with white. Very active reminiscent of the thickbilled flowerpecker. **Male**: Orange-red forecrown. **Female**: Lacks the orange on crown.
Call: Rendered as .. *spit* .. *spit* ...
Range: Resident. Goa southwards over the Western Ghats up to 2000 m.
Habitat: Evergreen and moist deciduous forests and secondary forests with bamboo.

804. Rufous Woodpecker *Cileus brachyurus*

Identification: 25 cm. A medium-sized chestnut-rufous woodpecker narrowly cross-barred with black on upperparts, wings and tail. Crest not distinct. Visits arboreal ant nests. **Male**: Distinguished by the presence of a crimson patch under the eyes. **Immature**: Barred on underparts.
Call: A loud nasal .. *kayn* .. *kayn* .. *kayn* .. rather similar to the common myna.
Range: Resident. Western Ghats up to 1000 m.
Habitat: Open and secondary evergreen forests with thickets, teak plantations and arecanut gardens.

808. Streakthroated Woodpecker *Picus xanthopygeus*
(Little Scalybellied Green Woodpecker)

Identification: 29 cm. A medium-sized green woodpecker with bright yellow back (in flight). Black-bordered white supercilium. Greenish-brown tail. Greyish underparts scalloped with black. Often on roadsides beside ant hills. **Male**: Crimson crown and crest. Black and orange nape. **Adult Female**: Black crown and crest. (*See* Lesser Yellownape, 816)
Call: A long drawn .. *peeer* .. similar in tone to that of lesser yellownape.
Range: Resident. Widespread on the Western Ghats up to 1800 m. Locally rare or absent. Common in the drier forests of Mudumalai and Bandipur Wildlife Sanctuaries.
Habitat: Deciduous, open forests and rubber plantations.

816. Lesser Yellownape *Picus chlorolophus*
(Small Yellownaped Woodpecker)

Identification: 27 cm. A medium-sized green woodpecker with maroon wings. Crimson eyebrow and yellow crest. Brownish underparts streaked and barred with white diagnostic (*See* streakthroated Woodpecker, 808). **Male**: Distinguished by the crimson forehead, supercilium and moustachial streak.
Call: A long drawn .. *cheeenk* .. or .. *peeer* .. confusable with the streakthroated woodpecker as well as the crested goshawk from a distance.
Range: Resident. Western Ghats up to 1800 m.
Habitat: Open evergreen and secondary forests with bamboo, plantations of rubber, coffee and also within hill settlements.

819, 820, 821. Blackrumped Flameback *Dinopium benghalense*
(Goldenbacked Woodpecker)

Identification: 29 cm. A medium-sized golden-orange, black and white wood-pecker with bright crimson crest. Chin, throat and breast black streaked / spotted with white. Rump black. **Male**: Crimson crown and crest. **Female**: Black (white-spotted) crown and crimson crest. The races differ subtly in colour. Race *benghalense* has a black throat streaked with white. Races *puncticolle* and *tehminae* have a black throat spotted with white; the latter, however, has a richer golden-olive-yellow back.

Call: Loud laugh .. *ke-ke-ke-kei-i-i-i* .. *i-i* .. somewhat similar to that of the whitethroated kingfisher.

Range: Resident. Entire Southwestern India up to 1200 m. The three races intergrade over their ranges. Race *benghalense* occurs from Maharashtra northwards and race *tehminae* southwards through Goa and Kerala. Race *puncticolle* occurs on the eastern side and not in Kerala.

Habitat: Evergreen forests (edges), secondary and deciduous forests, planta-tions, coastal coconut, beach and estuarine vegetation, orchards and groves in and around villages and towns.

825. Common Flameback *Dinopium javanense*
(Goldenbacked Threetoed Woodpecker)

Identification: 28 cm. A medium-sized golden-orange, black and white wood-pecker with crimson-red back visible in flight. White supercilium. Under-parts white scalloped with black. Single black band through chin and throat separate from similar greater flameback. **Male**: Crimson crown and crest. **Female**: Black crown and crest. In hand the absence of inner toe (smallest) diagnostic.

Call: Shriller and very different from blackrumped flameback .. *ti-ni-ni-ni-ni-ni* ...

Range: Resident. Goa and southwards on the Western Ghats up to 1700 m.

Habitat: Evergreen and secondary forests and plantations.

830. Whitebellied Woodpecker *Dryocopus javensis*
(Great Black Woodpecker)

Identification: 48 cm. A large black woodpecker with white rump, breast and abdomen. Crest crimson. **Male**: Crimson forehead, crown, crest and cheeks. **Female**: Crimson restricted to nape.

Call: A single loud .. *kek* ...

Range: Resident. Western Ghats up to 1200 m.

Habitat: Evergreen, deciduous, secondary forests and teak plantations.

847. Yellowcrowned Woodpecker *Dendrocopus mahrattensis*
(Yellowfronted Pied Woodpecker)

Identification: 18 cm. A small black and white woodpecker (spotted and streaked) with a yellow crown and red crest. Bright scarlet patch on abdomen. **Female**: Entire crown and crest yellow.
Call: Rendered as .. *click* .. *click* .. or .. *click-r-r-r* ...
Range: Resident. Entire Southwestern India up to 2000 m.
Habitat: Open secondary and deciduous forests.

852, 853. Browncapped Woodpecker *Dendrocopus nanus*
(Pygmy Woodpecker)

Identification: 13 cm. A very small uncrested brown and white woodpecker with a general plumage pattern as in the yellowcrowned woodpecker. Lack of any striking colours on plumage and small size diagnostic. **Male**: Distinguished from female by the crimson streak on either side of head. The races differ in intensity of colour. Race *cinereigula* has very dark upperparts including crown and finer streaking on the underparts as against race *hardwickii*.
Call: A shrill .. *click-r-r-r* .. *r* ...
Range: Resident. Western Ghats up to 1200 m. Race *cinereigula* occurs in the south from Coorg (southern Karnataka) through Kerala and Tamil Nadu.
Habitat: Evergreen, secondary and deciduous forests with scrub and thickets, teak plantations and taller trees bordering villages.

856. Heartspotted Woodpecker *Hemicircus canente*

Identification: 16 cm. A small, squat, black and creamy-white woodpecker with a very short tail and almost erect large crest. Black back and a buff band marked with black heart-shaped dots on wings diagnostic. **Female**: Differs from male in having the forehead and crown creamy-white (vs. black).
Call: Normal and on wings .. *click* .. *click* ... A trill .. *ti-ti-ti-ti* .. *ti* .. rather similar to that of the three-striped palm squirrel.
Range: Resident with local movements. Western Ghats up to 1300 m.
Habitat: Evergreen, secondary, deciduous and open forests along with the associated monocultures such as teak, and occasionally within suburban gardens.

858. Whitenaped Woodpecker *Chrysocolaptes festivus*
(Blackbacked Woodpecker)

Identification: 29 cm. A medium-sized golden-backed woodpecker identified by the distinct facial pattern, the white sides of hindneck and a large white 'V' on the middle of upper back. Lower back, rump and tail black. Underparts white scalloped with black. Five narrow black lines through sides of face, chin and throat. A woodpecker of drier vegetation. **Female**: Differs from male in having golden-yellow crown and crest (vs. crimson).

Call: Shrill call similar to that of the greater flameback. Rendered as .. *kwir-ri-rr-rr-rr-rr* ...

Range: Resident. Entire Southwestern India along the drier foothills.

Habitat: Deciduous forests and in village groves.

862. Greater Flameback *Chrysocolaptes lucidus*
(Larger Goldenbacked Woodpecker)

Identification: 31 cm. A largish golden-backed woodpecker identified by its size, crimson back and distinct facial pattern. Face five-striped as in whitenaped woodpecker (vs. single in common flameback). However, the first two meet on sides of face. **Female**: Differs from male in having black crown and crest (vs. crimson).

Call: Rather similar to and confusable with that of common flameback. Flight calls louder and diagnostic .. *ke-ke-ke* .. *ke-ke-ke* .. etc.

Range: Resident. Western Ghats up to 1800 m.

Habitat: Evergreen, secondary and deciduous forests, open thickets interspersed with taller trees along the moister zones and within groves in hill towns.

Family PITTIDAE

Small brightly coloured ground birds. Largish head. Short tail. Long legs. White patches on dark wings conspicuous in flight. Hopping about within shady undergrowth and groves. Sexes alike.

867. Indian Pitta *Pitta brachyura*

Identification: 19 cm. A small stub-tailed bright green, orange and crimson ground bird within bushes. Blue and black wings with white patches diagnostic in flight.

Call: A double whistle .. *piou-peeou* .. heard at dusk and dawn. Harsh alarm notes similar to that of the Oriental magpie robin.

Range: Resident with local movements in Southwestern India as far south as Uttara Kannada district (northern Karnataka). Winter visitor further south over the entire range up to 1700 m.

Habitat: Evergreen forests, secondary scrub and deciduous forests, plantations, urban groves and gardens.

Family ALAUDIDAE

Small squat birds, often with crests. Brownish and streaked plumage. On ground, in open country. Differ from rather similar pipits in heavier build and thicker beak. In hand the rear part of tarsus being scaled is diagnostic. Sexes generally alike.

872. Singing Lark *Mirafra cantillans* (Singing Bushlark)

Identification: 15 cm. A small hen-sparrow-like bird with weak fluttering flight. Wings rufous-chestnut. Hindneck not more distinctly streaked than the rest of upperparts. Outer tail feathers white. Song flight like that of skylark. (*See* Oriental Skylark, 907, 908)

Call: Lively as that of a oriental skylark. A greater repertoire and mimicry is, however, characteristic.

Range: Resident. Entire Southwestern India.

Habitat: Grassland, dry cultivation and fallow, scrub and estuarine waste.

874. Rufouswinged Lark *Mirafra assamica* (Madras Bushlark)

Identification: 15 cm. A small lark with rufous wings. Outer tail feathers brown. Flight weak when flushed. Characteristic parachuting flight display. Easily confused with Indian lark in the field. In hand, darker brown midstreak on each rufous primary feather is diagnostic.

Call: A shrill .. *see-see-see* ... When drifting on wings more lively .. *wisee-wisee-wisee* ...

Range: Resident. Entire Southwestern India.

Habitat: Scrub and fallow land, especially lateritic, and also beaches.

877. Indian Lark *Mirafra erythroptera* (Redwinged Bushlark)

Identification: 14 cm. A lark almost identical in all respects with the rufouswinged. The primaries are, however, entirely rufous with no black midstreak. Best identified in hand. (*See* Rufouswinged Lark, 874)

Call: Similar to Rufouswinged Lark.

Range: Resident. Widespread in Southwestern India except in Kerala.

Habitat: Open scrub, lateritic outcrops and cultivation.

878. Ashycrowned Sparrow-Lark *Eremopterix grisea*
(Ashycrowned Finch-Lark)

Identification: 13 cm. A very small squat sparrow-like grey-brown lark with thick crestless head and conical finch-like beak. **Male**: Ashy crown. White cheeks. Black chin and underparts.
Call: In flight a wheezy .. *seeeese* .. Song like that of Oriental skylark.
Range: Resident with local movements. Entire Southwestern India up to 1000 m.
Habitat: Open scrub, cultivation, dry wastelands along the coast, estuaries (salt pans), foothills and within urban limits.

882, 883. Rufoustailed Lark *Ammomanes phoenicurus*
(Rufoustailed Finch-Lark)

Identification: 16 cm. A medium-sized rufous-brown crestless lark with bright rufous tail and a black terminal band on tail. The race *testaceus* is more rufous than the race *phoenicurus*.
Call: Song described as pleasant and thrush-like. Another rendered as .. *tee-hoo* ...
Range: Resident. Widespread along the eastern foothills of Southwestern India except in Kerala. The race *testaceus* occurs in Karnataka and Tamil Nadu.
Habitat: Dry stony scrub and cultivation.

886. Greater Short-toed Lark *Calendrella brachydactyla*
(Rufous Short-toed Lark)

Identification: 15 cm. A small, pale coloured lark in large flocks in open semi-desert country. Outer tail feathers white. Smaller size and thicker conical beak separate from Oriental skylark. Lack of rufous on wings, longer blackish tail and streaks on breast (vs. spots) separate from other larks.
Call: Described as an occasional chirp.
Range: Winter visitor. The race *dukhunensis* is widespread over the foothills in Southwestern India. The less reddish race *longipennis* has, however, been recorded from Gujarat and northern Maharashtra.
Habitat: Open stony ground, cultivation, scrub and estuarine wastelands.

901. Malabar Lark *Galerida malabarica* (Malabar Crested Lark)

Identification: 15 cm. A small lark with a distinct erect blackish crest. Differs from *Galerida deva* which occurs in the plateau by larger size and heavier black streaking on breast. *Galerida cristata* is much larger and greyer and occurs in northern Gujarat.
Call: A shrill .. *ti-ee* ... A pleasant warbling song with a bit of mimicry.

Range: Endemic. Resident with local movements. Western Ghats up to 2000 m.
Habitat: Open scrub, grasslands, cultivation, beaches, estuarine and urban wastelands.

907, 908. Oriental Skylark *Alauda gulgula* (Small Skylark)

Identification: 17 cm. A medium-sized short-crested lark with white outer tail feathers. Lack of rufous on wings and stronger flight separate from Mirafra larks; larger size and darker colour from greater short-toed lark; shorter tail and squat build from pipits. Ascends the most among larks during song flights. The race *australis* is larger and darker in coloration than the race *gulgula*.
Call: A chirp when flushed. A lively though monotonous song similar to the Malabar lark and singing lark; probably more sustained. Certain phrases resemble the purple sunbird's.
Range: Resident with local movements. Entire Southwestern India. The race *australis* occurs in the Nilgiris and southward through Kerala and Tamil Nadu up to 1600 m.
Habitat: Open and wet grasslands, cultivation and fallow fields.

Family HIRUNDINIDAE

Small, slender aerial feeders with short beaks. Flight graceful. Differentiated from swifts by shorter and broader wings. Perch upright. Often in large flocks. Colonial mud nests characteristic. Sexes alike. **Martins:** short, nearly squared tails. **Swallows:** long, often deeply forked tails.

910. Sand Martin *Riparia riparia* (Collared Sand Martin)

Identification: 13 cm. A small smoky-brown nearly square-tailed martin with white underparts and a distinct smoky-brown band across breast. No white spots on tail.
Call: Rendered as .. *ret* .. or .. *brrit* ...
Range: Winter straggler. Sight records from southern Karnataka.
Habitat: Vicinity of water.

912. Plain Martin *Riparia paludicola* (Greythroated Sand Martin)

Identification: 12 cm. A small greyish-brown martin with no white spots on nearly squared tail. Head, wings and tail darker. Chin, throat and breast grey. Rest of underparts white. (*See* Sand Martin, 910)
Call: Described as similar to sand martin.
Range: Resident with local movements as far south as northern Karnataka. Stragglers sighted during winter in Uttara Kannada district.

Habitat: Estuaries and rivers.

913. Eurasian Crag Martin *Hirundo rupestris*

Identification: 14 cm. A sooty-brown martin with short nearly squared tail. Whitish underparts and black under tail-coverts diagnostic. Tail with round white spots conspicuous in flight.
Call: Rendered as a soft .. *chit-chit* ...
Range: Winter visitor. Widespread on the Western Ghats including the southern hills.
Habitat: Hillsides, cliffs and wet cultivation.

914. Dusky Crag Martin *Hirundo concolor*

Identification: 13 cm. A small sooty-brown martin separated from Eurasian crag martin by uniformly dark underparts. Chin, throat and foreneck finely streaked with black.
Call: Described as similar to Eurasian crag martin.
Range: Resident with local movements. Entire Southwestern India up to 1800 m.
Habitat: Exposed hillsides and cliffs overhanging rivers, waterfalls, under bridges, dams, forts and old buildings.

916, 917. Barn Swallow *Hirundo rustica* (Common Swallow)

Identification: 18 cm. A medium-sized glossy blue and dirty white swallow with a deeply forked tail. Chestnut forehead, chin and throat. Black border separates throat and breast. White spots on tail conspicuous in flight and diagnostic. Large gatherings on overhead wires during early and late winter. **Immature**: Overall less bright. Browner. Outer tail feathers shorter making the tail appear squarish. The race *rustica* has pale rufous underparts and a complete pectoral band. The race *gutturalis* has white underparts with an incomplete pectoral band.
Call: Twitters .. *which-which* ...
Range: Winter visitor. Entire Southwestern India.
Habitat: Open scrub, cultivation, marshes, beaches and in urban neighbourhoods.

919. Hill Swallow *Hirundo domicola* (House Swallow)

Identification: 13 cm. A small glossy black and chestnut swallow with deeply forked tail. Grey abdomen and undertail. Lack of a pectoral band and dark underwings separate from common swallow. Deeply forked tail distinguishes from dusky crag martin. Hills.
Call: .. *which-which* .. as barn swallow.

Range: Resident. Southwards from Coorg (southern Karnataka) over the Western Ghats from about 700 m up to over 2000 m.

Habitat: Grassy slopes, estates and in old buildings and ruins.

921. Wiretailed Swallow *Hirundo smithii*

Identification: 14 cm (10–13 cm wires). A small glossy blue and glistening white swallow with a chestnut cap. Readily identified by the long wires (outermost tail feathers; shorter in females). In flight, glistening white underparts contrasting with the blue upperparts diagnostic. **Immature**: Sooty above with traces of blue. No wires on tail.

Call: .. *chit-chit* .. or .. *which-which* .. while flying about.

Range: Resident with local movements. Entire Southwestern India up to about 2000 m.

Habitat: Open scrub, cultivation, marshes of all sorts including estuaries and rivers, waterfalls and in urban neighbourhoods.

922. Streakthroated Swallow *Hirundo fluvicola* (Cliff Swallow)

Identification: 12 cm. A small nearly square-tailed swallow. Glossy blue above with dull chestnut forehead and crown. Rump pale brown. Dirty white below streaked with black on chin, throat and breast. In flight, contrasting paler rump and short tail confusable with northern house martin in poor light.

Call: Described as a sharp .. *trr-trr* .. in flight.

Range: Resident with local movements. Widespread in Southwestern India as far south as Nilgiris (Silent Valley) up to about 1000 m.

Habitat: Open forests, cultivation, grasslands and marshes including estuaries.

925, 927. Redrumped Swallow *Hirundo daurica*

Identification: 19 cm. A fairly big brownish swallow with deeply forked tail. Deep blue above with a chestnut supercilium, sides of face and collar on hindneck. Chestnut rump conspicuous in flight. Whitish underparts streaked with blackish. Black under tail-coverts. Paler rump and lack of white spots on tail separate from barn swallow in flight. Large gatherings on overhead wires (*See* Steakthroated Swallow, 922). The race *nipalensis* differs from the race *erythropygia* in having a paler rump and more boldly streaked underside.

Call: Rendered as .. *cheer* .. or .. *queenk* ...

Range: Resident. Entire Southwestern India up to 1600 m. The race *nipalensis* is a winter visitor.

Habitat: Open forests, grassland, marshes, cultivation, beaches and urban neighbourhoods including buildings.

930. Northern House Martin *Delichon urbica*

Identification: 15 cm. A small black and white swallow with a shallow fork on tail. Glistening white rump conspicuous in flight. Fully white underparts and broader wings separate from little swifts. Legs and feet fully covered with white feathers diagnostic at close range.

Range: Winter visitor and on passage over the Western Ghats up to at least 1000 m.

Habitat: Open forests, estuaries and urban neighbourhoods.

Family LANIIDAE

Medium-sized birds with strong beaks hooked at the tip. Proportionately large and rounded heads. Longish tails. Black band through eyes. Take prey from ground. Pin prey to thorns, hence the name 'Butcher Birds'. Perch low singly and idly. Harsh calls. Sexes alike.

933. Northern Shrike *Lanius exubitor* (Grey Shrike)

Identification: 25 cm. A large silvery-grey and white bird with heavy hooked bill and longish graduated tail. Black band through eyes diagnostic. In flight white mirror on black wings and white outer tail conspicuous. Singly on exposed perches and on overhead wires. Race *lahtora* confusable with no other shrike except the race *caniceps* of the longtailed shrike. **Immature**: Paler spots and bars on plumage.

Call: Harsh .. *che-che-che*... etc., when alarmed. Another rendered as .. *kwi-ric.., kwi-rick*...

Range: Resident with local movements. Eastern sides of Southwestern India as far south as Belgaum (northern Karnataka).

Habitat: Dry deciduous forests, thorny scrub and cultivation.

940. Baybacked Shrike *Lanius vittatus*

Identification: 18 cm. A small brightly coloured shrike perched low on exposed bushes. Head and underparts largely white with a prominent black band through eyes. Black beak. Chestnut-maroon back. Black graduated tail. In flight white rump, white mirror on black wings, white outer tail feathers and maroon back diagnostic. Smaller size and much brighter colour on back separate from longtailed shrike. **Immature**: Rufous tail and no wing mirrors. Confusable with rufoustailed shrike. Grey rump however diagnostic.

Call: A squeal rendered as .. *chee-urr* ...

Range: Resident with local movements. Entire Southwestern India along the drier eastern foothills.

Habitat: Open forest, thorny scrub, cultivation and urban wastelands.

943. Rufoustailed Shrike *Lanius isabellinus* (Pale Brown Shrike)

Identification: 17 cm. A small pale brownish and rufous shrike with a white supercilium and without a distinctly white wing mirror. Confusable with brown shrike. However paler upperparts and darker rufous-red tail diagnostic. Female and Immature: Lack the dark band through eyes. Variable amount of barring on upper and / or lower plumage.
Range: Winter visitor. Gujarat and Maharashtra south to Mumbai.
Habitat: Dry deciduous thorn and scrub.

946, 947. Longtailed Shrike *Lanius schach* (Rufousbacked Shrike)

Identification: 25 cm. A large grey, white and rufous shrike with black beak, forehead and band through eyes. White mirrors on wings. Black and rufous tail. Rufous back, rump and abdomen diagnostic. The grey-backed race *caniceps* is confusable with northern shrike. Traces of rufous on tail and abdomen and a bright rufous rump identify the grey-backed race. Prefers more humid habitats. Immature: Duller and lack white wing mirror. Larger size and greyer upperparts separate from rufoustailed and brown shrikes.
Call: Harsh squeaks rendered as .. *gerlek-gerlek-yaon-yaon* ... A good mimic of other species.
Range: Resident with local movements in Southwestern India. The race *erythronotus* is a winter visitor south of Belgaum (northern Karnataka). The race *caniceps* occurs over the entire range up to 2000 m (Nilgiris).
Habitat: Open forests, clearings in evergreen forests, grass and scrub, plantations, cultivation, edges of marshes including estuaries, beaches and urban gardens.

949. Brown Shrike *Lanius cristatus*

Identification: 19 cm. A small dull-brown and whitish shrike with a dark band through eyes and no white mirror on wings. Overall darker and reddish upperparts and rufous (not reddish) tail separate from rufoustailed shrike. Some adults have barring on breast and flanks as the immature. Low bushes. Characteristic loud calls.
Call: Harsh .. *che-che-che* ...
Range: Winter visitor. Entire Southwestern India.
Habitat: Forest clearings, open forests with thorn and scrub, plantations, cultivation, edges of marshes including estuaries, beaches, urban gardens and wastelands.

Family ORIOLIDAE

Medium-sized birds with bright golden-yellow plumage. Red beaks. Longish wings. Rich quality of songs. Sexes differ.

NOTE: Sibley and Monroe (1990) treat Orioles, Minivets, Cuckoo-Shrikes, Wood-Shrikes and Flycatcher-Shrikes together with Crows, Drongos, Wood Swallows, Fantail, Paradise and Monarch Flycatchers and Ioras as belonging to family CORVIDAE.

953. Eurasian Golden Oriole *Oriolus oriolus* (Golden Oriole)

Identification: 25 cm. A bright golden-yellow bird with red beak, black wings and tail. Black line through eyes. **Female** and **Immature**: Dull green upperparts and whitish (brown-streaked) underparts. No black on plumage.
Call: A harsh .. *chearrrr* ... A musical .. *pee-lo-lo* ...
Range: Winter visitor. Entire Southwestern India up to 1800 m.
Habitat: Evergreen forests and associated stages of degradation, deciduous forests, monocultures including eucalyptus and rubber, cultivation, groves, orchards and urban gardens.

954. Blacknaped Oriole *Oriolus chinensis*

Identification: 25 cm. An unmistakable golden-yellow oriole with black wings and tail. Separated from Eurasian golden oriole by the black bands through eyes extending to meet on nape. **Female**: Like male except for the greener plumage.
Call: Rendered as .. *wheeow* .. and .. *chuck-tarry-you* ...
Range: Winter visitor. Western Ghats up to 1000 m.
Habitat: Evergreen and deciduous forests including their secondary stages and monocultures.

959. Blackhooded Oriole *Oriolus xanthornus* (Blackheaded Oriole)

Identification: 25 cm. A bright golden-yellow oriole with black wings and tail readily identified by its black head and neck. **Immature**: Yellow on head and forehead. White throat. Breast streaked with black.
Call: A loud .. *proink* ... Song richer than that of Eurasian golden oriole.
Range: Resident. Western Ghats up to 1700 m.
Habitat: Evergreen and deciduous forests including their secondary stages, monocultures, orchards and urban groves.

Family DICRURIDAE

Medium to large-sized birds. Black plumage. Tail deeply forked or with ornaments. Slender slightly hooked beaks. Active and noisy. Remarkable vocal mimicry.

962, 963. Black Drongo *Dicrurus macrocercus*

Identification: 31 cm. A medium-sized glossy black bird readily identified by its long fish-like tail. Brownish flight feathers. Separated from other drongos by a white spot at base of beak and a preference for open country.
Call: A variety of loud scolding sounds; most common being .. *ti-hee* .. similar to that of shikra. Also .. *pree* .. *preewick-wick* ...
Range: Resident. Entire Southwestern India up to 2100 m. The slightly larger race *albirictus* straggles into Maharashtra in winter.
Habitat: Open country with cultivation, scrub and marshes, estuaries, beaches and urban wastelands.

965. Ashy Drongo *Dicrurus leucophaeus* (Grey Drongo)

Identification: 30 cm. A slim grey-black long-tailed drongo with conspicuous red eyes. Separated from black drongo by duller plumage, slimmer build with deeper fork on tail, a greater preference for wooded habitats and a more arboreal habit. Flies high over tree canopy. Flocks on flowering silk cotton, coral and eucalyptus trees.
Call: Like that of black drongo including a lot of mimicry. A common call rendered as .. *drang-gip* .. *gip* .. *gip-drang* ...
Range: Winter visitor. Entire Southwestern India up to 1500 m.
Habitat: Evergreen forests (edges and canopy), deciduous forests, all secondary forests and monocultures, orchards, groves and urban gardens.

967. Whitebellied Drongo *Dicrurus caerulescens*

Identification: 24 cm. A small, slender drongo readily identified by its white belly and undertail.
Call: Harsh calls. Song includes a variety of mimicry. Generally shriller and softer than all other drongos.
Range: Resident with local movements. Entire Southwestern India up to 1500 m.
Habitat: Secondary and deciduous forests, thickets with taller trees interspersed, monocultures and urban groves.

971. Bronzed Drongo *Dicrurus aeneus*

Identification: 24 cm. A small bronze-green and blue glossed black drongo with slightly forked tail. Flocks in forests.
Call: Loud .. *proi-preek-preek-preek* .. very similar to the greater racket-tailed drongo from a distance. A good mimic.
Range: Resident with local movements especially during the rains. Western Ghats south of Khandala (Maharashtra) up to 2000 m.
Habitat: Evergreen forests, secondary and moist deciduous forests, monocultures and hill orchards.

973. Haircrested Drongo *Dicrurus hottentottus*

Identification: 31 cm. A medium-sized glossy black drongo with longish curved beak. Characteristically upward twisted outermost tail feathers diagnostic. At close range hair-like feathers from forehead falling over crown visible. Flocks on flowering silk cotton and coral trees.
Call: Loud and noisy with a lot of mimicry. Normally a sharp .. *twee* ...
Range: Resident with local movements. Entire Southwestern India (straggling north as far as Kutch) up to 1400 m.
Habitat: Open secondary and deciduous forests with scrub and thicket, monocultures (especially eucalyptus), and within suburbs.

977. Greater Racket-tailed Drongo *Dicrurus paradiseus*

Identification: 35 cm (+ 35 cm tail). Large glossy black drongo with a prominent backward curving crest and two long spatula-tipped wiry extensions from tail. Large size and crest are diagnostic when rackets are lost in moulting. Noisy.
Call: Harsh .. *broik* .. *broik* .. *broik* ... A characteristic .. *pr-pr-proik-preek-preek* .. similar to that of bronzed drongo but much louder. A variety of sounds with a fantastic repertoire of mimicry.
Range: Resident. Western Ghats up to 1500 m.
Habitat: Evergreen, deciduous and secondary forests, monocultures, open scrub and thicket in cleared forests and outskirts of towns.

Family ARTAMIDAE

Small stumpy birds with thick conical beaks. Graceful swallow-like flight. Broader wings and shorter squared tails distinguish from swallows. Sexes alike.

982. Ashy Wood Swallow *Artamus fuscus* (Ashy Swallow Shrike)

Identification: 19 cm. Small. Slaty-grey long-winged bird with paler rump and underparts. Thick beak and wing-tips reaching end of tail diagnostic at rest. Swallow-like flight and characteristic calls are further pointers. Pairs or small flocks on exposed branches and overhead wires.
Call: Harsh .. *chek-chek-chek* .. at rest and in flight.
Range: Resident with local movements. Entire Southwestern India up to 2100 m.
Habitat: Evergreen forests (edges and gaps), secondary and deciduous forests, monocultures especially of palms, scrub, cultivation, marshes, around estuaries and beaches and within towns.

Family STURNIDAE

Medium-sized to fairly big birds with pointed beaks and short tails. Arboreal though freely forage on ground. Often follow cattle. Hole-nesters. Gregarious and noisy. Sexes alike.

987, 988. Chestnut-tailed Myna *Sturnus malabaricus* (Greyheaded Myna)

Identification: 21 cm. Small silvery-grey and rufous myna in flocks on flowering coral and silk cotton trees. Greyish-white head and rufous underparts diagnostic. The **Male** of race *blythii* has pure white head, neck and breast and much darker underparts. **Female** *blythii* is rather similar to race *malabaricus* except that it shows more silvery-grey on head and neck, pure white on chin and throat and paler rufous on underparts.
Call: A soft .. *ke-kik* .. as the birds fly and when at flowers.
Range: Resident with local movements in Southwestern India up to 1200 m. The race *blythii* is resident on the Western Ghats from Belgaum southwards straggling further north to Mumbai.
Habitat: Evergreen forests (edges and gaps), secondary and deciduous forests, open scrub interspersed with taller trees, cultivation, monocultures including eucalyptus and urban groves.

994. Brahminy Starling *Sturnus pagodarum* (Blackheaded Myna)

Identification: 22 cm. A small grey and reddish-brown myna with a black cap and crest. Black head and black wings without a white mirror and white tips to brown tail diagnostic in flight. **Immature**: Black on head does not extend beyond nape, appearing as if the birds have had a haircut.
Call: Creaking sounds, .. *creoo-cree-creek-creek* ...
Range: Resident. Entire Southwestern India.

Habitat: Secondary and open deciduous forests, cultivation, scrub and urban gardens.

996. Rosy Starling *Sturnus roseus* (Rosy Pastor)

Identification: 23 cm. A medium-sized rosy-pink starling with shiny black head, crest, neck, breast, wings and tail. Large noisy flocks in fields and on flowering silk cotton trees. **Immature**: Sandy-brown with darker head, wings and tail. Crestless.
Call: Noisy chattering .. *kit-kit-kit* ...
Range: Winter visitor. Entire Southwestern India.
Habitat: Open cultivation, scrub, estuarine groves, along beaches and within urban campuses.

997. Common Starling *Sturnus vulgaris*

Identification: 20 cm. Small glossy green-purple and black starling with white spots on head, neck and breast. Bright yellow beak diagnostic. Flocks.
Call: Described as creaky chatterings and musical whistles.
Range: Winter visitor. Gujarat.
Habitat: Cultivation, outskirts of towns and along marshes.

1002. Asian Pied Starling *Sturnus contra* (Pied Myna)

Identification: 23 cm. Medium-sized, slim black and white myna. Red skin around eyes, orange and yellow beak, white face, shoulders and rump diagnostic. **Immature**: Duller. Breast streaked with brown.
Call: Described as high-pitched musical notes.
Range: Stragglers to Mumbai (Borivali).
Habitat: Open country, cultivation, meadows and suburbs.

1006. Common Myna *Acridotheres tristis*

Identification: 23 cm. A familiar medium-sized brown myna with black head, neck and breast. Bright yellow beak, legs and patch around eyes and in flight, white mirrors on black wings, white under tail-coverts and white tips to dark tail diagnostic. Pairs or small flocks in fields often following cattle. **Immature**: Browner with patch around eyes smaller and paler.
Call: Noisy .. *kay-kay-kay* ... A variety of others including .. *trok-trok-trok* .. *treek* .. *treek* .. and what is rendered as .. *radio* .. *radio* ...
Range: Resident. Entire Southwestern India. Less common in the hills.
Habitat: Open and secondary forests, scrub, monoculture, cultivation, marshes, estuaries, beaches and urban homesteads.

1008. Bank Myna *Acridotheres ginginianus*

Identification: 21 cm. Small myna. Rather similar to common myna except for the blue-grey plumage and reddish bare patch around eye. A tuft of upright feathers at base of beak. In flight buffy wing mirror and tips to tail diagnostic. **Immature**: Browner with flecked underparts. Grey iris (vs. red in adults).
Call: Described as similar but softer than common myna.
Range: Resident with local movements. South as far as Mumbai.
Habitat: Urban neighbourhoods and cultivation.

1010. Jungle Myna *Acridotheres fuscus*

Identification: 23 cm. A medium-sized myna distinguished from common myna by slenderer build, slaty-grey plumage and lack of a bare patch around eyes. Strikingly white eyes on black head and prominent erect tuft at base of yellow beak diagnostic. In flight, white mirrors on wings conspicuous. Erect tuft on base of beak often clearly visible. Attend grazing buffaloes and cattle.
Call: Rather different from common myna. Lower in tone and harsher .. *proik-proik* ...
Range: Resident. Entire Southwestern India up to over 2400 m.
Habitat: Open secondary and deciduous forests, scrub, monocultures, cultivation, marshes, estuaries and urban gardens.

1016. Hill Myna *Gracula religiosa*

Identification: 25 cm. A large glossy black myna with orange-red beak and fleshy wattles on sides of head and nape. In flight, white mirrors against jet-black body and wings and a 'whizz', diagnostic. Arboreal. Pairs or small flocks flying high.
Call: Loud .. *awk* .. *wheeeow* ... A variety of squeaks. A good mimic in captivity.
Range: Resident with local movements. Western Ghats south of Maharashtra up to 1700 m. Naturalized local populations in Mumbai are probably escapees from pet shops.
Habitat: Evergreen, secondary and deciduous forests, grasslands interspersed with trees, around hill settlements and in a variety of hill cultivations and tree plantations.

Family CORVIDAE

Large birds with powerful beaks. Aggressive and often predatory. Harsh calls. Sexes alike. **Tree Pies:** slender, long-tailed colourful birds. Pairs in forests and wooded habitats. **Crows:** black, urban birds. Gregarious.

1031, 1033, 1034. Rufous Tree Pie *Dendrocitta vagabunda*
(Indian Tree Pie)

Identification: 25–30 cm. A slender, long-tailed black and rufous koel-like bird. Silvery-grey tail, rufous body and black and white-grey wings diagnostic in flight. Harsh calls. Pairs. The race *parvula* is the darkest and smallest. The races *vernayi* and *pallida* are paler, the latter being the larger and darker of the two.

Call: Harsh .. *ka-ka-ka-ka-ka* .. *ka* ... Another .. *uk-ru-ni* .. also rendered as .. *bob-o-link* ...

Range: Resident and found up to 2000 m. The race *parvula* is the most widespread in Southwestern India being replaced by the races *pallida* and *vernayi* in Maharashtra and the Nilgiris–Palnis respectively.

Habitat: Open, secondary and deciduous forests, monocultures including eucalyptus and rubber, estates, orchards, coastal coconut and urban gardens.

1036. Whitebellied Tree Pie *Dendrocitta leucogastra*

Identification: 50 cm. An exceptionally long-tailed tree pie. Black, white and chestnut plumage. In flight, white mirrors on black wings and long black and white graduated tail diagnostic. Pairs.

Call: Loud .. nasal .. *koyn-koyn-koyn* .. and .. *proink* .. confusable with the greater racket-tailed drongo and blackhooded oriole.

Range: Resident. Endemic. Western Ghats south from Goa through Karnataka, Tamil Nadu and Kerala up to 1500 m.

Habitat: Evergreen forests and their secondary stages, moist deciduous forests and in the vicinity of human settlements in the hill stations.

1049, 1050. House Crow *Corvus splendens*

Identification: 43 cm. A small familiar crow with greyish neck and breast. The darker race *protegatus* lacks the contrasting grey on neck and breast.

Call: Familiar .. *kaa-kaa-kaa* ... Also a variety of chuckles.

Range: Resident. Entire Southwestern India up to 2100 m. Race *splendens* is replaced by race *protegatus* in Kerala south of the Palghat gap.

Habitat: Open scrub, cultivation, groves, mangrove, marshes and beaches; always close to human habitation.

1057. Largebilled Crow *Corvus macrorhynchos* (Jungle Crow)

Identification: 48 cm. A large glossy black crow with long heavy beak. A diminutive version of the raven.
Call: Deep and hoarse .. *crow* ...
Range: Resident. Entire Southwestern India up to over 2300 m.
Habitat: Evergreen forests (edges), secondary and deciduous forests, monocultures, scrub and grasslands, cultivation, marshes, beaches and urban gardens not very far from human settlements.

Family BOMBYCILLIDAE

Hypocolius: Short and stout legs and beaks. Short wings and long graduated tails.

NOTE: Sibley and Monroe (1990) treat *Hypocolius* as belonging to family HYPOCOLIIDAE.

1063. Grey Hypocolius *Hypocolius ampelinus*

Identification: 25 cm. A grey and black shrike-like bird with a black band through eyes to nape and black erectile crest. Black flight feathers with white tips and longish blue-grey tail with black terminal band diagnostic.
Female: Cream underparts. Lacks black band through eyes. Flight feathers grey-brown with black and white tips.
Range: Straggler into Gujarat and Maharashtra (Raigad district).
Habitat: Semi-desert and open scrub jungle.

Family CAMPEPHAGIDAE

Small to medium-sized insectivorous birds. Hooked beaks and slender upright build. Pairs or small flocks. Arboreal. **Flycatcher-Shrikes** and **Minivets**: small birds with longish, narrow tails. Brightly coloured: combinations of white, yellow and orange-red with black. Very active. Sexes strikingly different. **Wood-Shrikes** and **Cuckoo-Shrikes**: small to medium-sized rather long-winged grey-brown birds with short, broad tails. Thick, shrike-like beaks and broad black line through eyes. Often with bars on plumage. Sexes more or less alike.

1065, 1066. Barwinged Flycatcher-Shrike *Hemipus picatus*
(Pied Flycatcher-Shrike)

Identification: 14 cm. Small active black and white forest birds. Glossy black head and back. White hindcollar, rump and underparts. Wings and tail

black and white. Hunchbacked posture when perched is diagnostic. In flight
white patches on wings striking. Pairs or small flocks in the canopy.
Female: Browner upperparts. The races are barely different except in
colour of eyes which is orange-yellowish-brown in race *picatus* and brown
in race *leggei*.

Call: Shrill .. *whirriri* .. *whirriri* .. *whirriri* ...

Range: Resident with local movements. Western Ghats up to the highest
hills (over 2000 m). The race *leggei* represents the species in southern
Kerala.

Habitat: Evergreen forests, secondary and deciduous forests, teak and rubber
plantations and in estates in the neighbourhood of human settlements.

1068. Large Wood-Shrike *Tephrodornis gularis* (Malabar Wood-Shrike)

Identification: 23 cm. A medium-sized short square-tailed grey-brown shrike
of the forests. Grey crown, black band through eyes and brownish outer
tail feathers. White rump (finely barred), paler tips to tail and whitish underparts contrasting with darker crown diagnostic in flight. Small flocks.
Female: Browner crown and eye band. **Immature**: Mottled on head and
back.

Call: Harsh chatter .. *kra-cha-cha-cha-cha-cha* .. and .. *witoo* .. *witoo* .. *witoo*.

Range: Resident. Western Ghats up to 1800 m.

Habitat: Evergreen forests, secondary and moist deciduous forests and associated monocultures such as teak.

1070. Common Wood-Shrike *Tephrodornis pondicerianus*

Identification: 16 cm. A small brownish shrike-like bird of open woodlands.
Differentiated from large wood-shrike by smaller size, browner plumage,
a prominent pale supercilium above darker eye band and white outer tail
feathers. Shorter tail, more active arboreal foraging and flocks separate
from rather similar brown shrike.

Call: A feeble .. *weet-weet* .. and .. *wee-wit-wit-wit-wit-wit* .. confusable with
the common iora.

Range: Resident with local movements. Entire Southwestern India up to
1000 m.

Habitat: Open dry forests, monocultures of teak and eucalyptus, scrub, cultivation, orchards and suburban campuses.

1072. Large Cuckoo-Shrike *Coracina macei*

Identification: 28 cm. A large grey and white cuckoo-like forest bird readily
identified by its piercing call and habit of flicking wings on alighting.
Blackish mask and unbarred grey and white underside identifies the male.
Pairs on canopy.

Call: A sharp .. *ti-e-ee* ...
Range: Resident with local movements. Western Ghats up to 1200 m.
Habitat: Evergreen forests, secondary and deciduous forests, monocultures and urban campuses.

1077. Blackwinged Cuckoo-Shrike *Coracina melaschistos*
(Dark Grey Cuckoo-Shrike)

Identification: 22 cm. A medium-sized bluish-grey bird with a dark band through eyes and white tips to tail. Rather similar in appearance to adult greybellied cuckoo. **Female**: Paler grey with a white patch on underwing.
Call: Rendered as .. *tweet-tweet-tweer* ...
Range: Winter straggler known from Savantwadi (Maharashtra) and Londa (northern Karnataka) up to 600 m.
Habitat: Open forests.

1079. Blackheaded Cuckoo-Shrike *Coracina melanoptera*

Identification: 19 cm. A small pale grey and white short-tailed bird with black head, wings and tail. White tips to tail diagnostic. **Female and Immature**: Rather cuckoo-like. Grey head. Underparts barred with grey except on abdomen and vent.
Call: Normally silent in winter when most often seen. Song rendered as .. *pit-pit-pit* .. and .. *wheet-wheet-wheet* ...
Range: Resident with local movements. Entire Southwestern India up to 2100 m.
Habitat: Open forests of evergreen and deciduous types, thorny scrub, monocultures, groves and urban gardens.

1081. Scarlet Minivet *Pericrocotus flammeus* (Orange Minivet)

Identification: 20 cm. A slender scarlet-orange and black long-tailed bird of the forest canopy. Orange patches on black wings and orange outer tail feathers diagnostic in flight. Pairs or small flocks. **Female**: Largely golden-yellow (rather oriole-like). Black parts of male replaced by blue-grey or slate.
Call: Rich .. *twitwee-twitwee-twitwee* ... Especially when a flock takes off from the canopy.
Range: Resident. Western Ghats up to over 2000 m.
Habitat: Evergreen and deciduous forests including their stages of degradation, monocultures including teak and rubber, groves and suburban gardens.

1085. Longtailed Minivet *Pericrocotus ethologus*

Identification: 18 cm. A black and scarlet minivet difficult to distinguish in the field from scarlet minivet. However, smaller size, longer tail and lack of scarlet patches on innermost secondaries (visible even on closed wings) diagnostic. **Female:** Yellow and grey as in female scarlet minivet. Differs the same way as the male does from scarlet minivet. Lacks yellow patches on innermost secondaries.
Range: Winter straggler. A single record from Sangola (Maharashtra).
Habitat: Groves and wooded campuses.

1089. Rosy Minivet *Pericrocotus roseus*

Identification: 18 cm. A rosy-red and grey-brown minivet. Pink patches on dark blackish wings and rosy underparts diagnostic in flight. **Female:** Grey and yellow as other female minivets. Olive-yellow (vs. golden-yellow) rump diagnostic.
Call: Rendered as .. *whiririri-whiririri-whiririri* .. similar to the barwinged flycatcher-shrike.
Range: Winter straggler into Maharashtra, Goa and Kerala.
Habitat: Forests.

1089a. Ashy Minivet *Pericrocotus divaricatus*

Identification: 18 cm. A grey, black and white minivet with no trace of red or yellow on plumage. White wing patch and outer tail feathers conspicuous in flight. White forehead, greyer back, lack of a white hindcollar and larger size separate from barwinged flycatcher-shrike. **Female:** Lacks black on upperparts.
Call: Described as harsh and resembling that of shrikes.
Range: Winter straggler to Mumbai.
Habitat: Forest.

1093, 1094. Small Minivet *Pericrocotus cinnamomeus*

Identification: 15 cm. A small long-tailed grey, orange, yellow and whitish minivet. Orange patches on wings and call diagnostic in flight. Small flocks in the canopy. **Female:** Duller. Lacks the black throat and orange breast of male. Whitish-yellow underparts. The races differ primarily in the colour of throat, being darker and black in race *malabaricus*.
Call: Feeble .. *see-see-see* ...
Range: Resident. Entire Southwestern India up to 1000 m. Race *malabaricus* replaces race *cinnamomeus* southwards from Belgaum–Goa.
Habitat: Open secondary forests, monocultures, groves and urban gardens.

1096. Whitebellied Minivet *Pericrocotus erythropygius*

Identification: 15 cm. A small black and white minivet distinguished from other minivets by the pied coloration, and from the barwinged flycatcher-shrike by orange rump and breast. White wing-patch forms a 'V' on back when closed and is diagnostic. **Female**: Greyer upperparts.
Call: Described as similar to a wagtail's .. *tseep-tseep* ...
Range: Resident. Locally known from Gujarat, Belgaum (northern Karnataka) and Nilgiris.
Habitat: Grassy forests and scrub.

Family IRENIDAE

Small to medium-sized birds foraging within foliage. Mostly green plumage. Occasionally yellow or blue. Mellow calls and songs. Males usually brighter.

NOTE: Sibley and Monroe (1990) treat Ioras under family CORVIDAE.

1100, 1101. Common Iora *Aegithina tiphia*

Identification: 14 cm. Small greenish-yellow and black birds among foliage. Pairs or small flocks. **Adult Male breeding**: Black above. Yellow below. Two white bars on wings diagnostic. **Non-breeding Male** and **Female**: Greenish overall with brownish wings and the white bars. Black tail identifies the male.
Call: A long drawn .. *wee-e-e-e-u* ... Also .. *chee-chit-chit-chit* .. rather similar to the common wood-shrike. Another rendered as .. *if-you-please* .. similar to the puffthroated babbler.
Range: Resident. Entire Southwestern India up to 1500 m. The race *multicolor* replaces the race *deignani* south of the Palghat gap.
Habitat: Edges of evergreen forests, secondary and deciduous forests, monocultures, cultivations, groves, mangroves and urban gardens.

1104, 1105. Goldenfronted Leafbird *Chloropsis aurifrons*
        (Goldfronted Chloropsis)

Identification: 19 cm. A medium-sized leaf-green bird readily identified by its black beak, chin and throat and orange-red forehead. Sexes alike. Often concealed among dense foliage mimicking a wide variety of bird calls.
Call: A drongo-like .. *seek-seek* ... Others commonly heard are .. *preeki-preeki* .. *preek-preek-preek* .. and .. *treeek-pik-pik-pik* ... Mimicry includes more forest birds' calls and songs.
Range: Resident. Entire Southwestern India up to 1800 m (race *frontalis*). The race *insularis* is found south of the Palghat gap.

Habitat: Evergreen, secondary and deciduous forests, monocultures and urban gardens.

1107. Bluewinged Leafbird *Chloropsis cochinchinensis*
(Jerdon's Chloropsis)

Identification: 18 cm. A leaf-green medium-sized bird rather similar to the goldenfronted leafbird except for the absence of orange-red forehead. Black on chin and throat less extensive and bordered with yellow. Within foliage in opener forests and gardens. **Female**: Black on chin and throat replaced by bluish.
Call: Rather similar to goldenfronted leafbird. However, with some practice can be separated due to its smaller repertoire of mimicry and most frequently imitating the bulbuls.
Range: Resident. Entire Southwestern India up to 1200 m.
Habitat: Open secondary and deciduous forests, monocultures, groves and urban gardens.

1109. Asian Fairy Bluebird *Irena puella*

Identification: 27 cm. A fairly large flashing blue and black bird of the forest with crimson bead-like eyes. Pairs in mixed flocks. Characteristic calls. **Female**: Dull blue appearing greenish-grey in poor light.
Call: .. *pee-pik* .. *pee-pik* ... While taking to wings .. *pi-pik-pi-pik-pi-pik* ..
Range: Resident with local movements. Western Ghats south from Ratnagiri (Maharashtra) up to 1800 m.
Habitat: Evergreen, secondary and moist deciduous forests, monocultures including teak, rubber and arecanut and also suburban groves.

Family PYCNONOTIDAE

Medium-sized active birds in pairs or flocks. Often possess crests. Colour ranges from black to yellow. Noisy. In gardens, cultivation and tall forests. Sexes alike.

1114. Greyheaded Bulbul *Pycnonotus priocephalus*

Identification: 19 cm. An olive-green forest bulbul with a grey crestless head. Yellowish underparts. Whitish eyes. Dark mottling on rump.
Call: A sharp .. *prink* ...
Range: Resident. Endemic. Western Ghats south from Goa up to 1800 m.
Habitat: Evergreen forests and moist secondary open forests with *Eupatorium, Lantana*, etc., along the coastal foothills.

1116. Blackcrested Bulbul *Pycnonotus melanicterus* (Rubythroated Bulbul)

Identification: 18 cm. A small bright yellow and olive uncrested forest bulbul with a black head and ruby-red chin and throat. White eyes conspicuous.

NOTE: There are a couple of unconfirmed sight records of the race *flaviventris* with full yellow plumage, black head and sharp upright crest from Goa and Kalakkad hills. This race has an otherwise easterly range and is not known southwest of Andhra Pradesh.

Call: Musical and rather similar to redwhiskered bulbul .. *pik-pili-lou* .. *pi-pik* .. *pi-pik...*

Range: Resident. Western Ghats south from Goa up to 1200 m (race *gularis*).

Habitat: Evergreen forests, secondary moist forests with scrub and thickets and monocultures of teak.

1120. Redwhiskered Bulbul *Pycnonotus jocosus*

Identification: 20 cm. A lively brown and white crested bulbul of gardens and woodlands. Upright, sharp black crest, red ears, a blackish incomplete necklace across white breast and red vent diagnostic. **Immature**: Lacks the red whiskers. Yellowish vent.

Call: Familiar .. *pik-pikloo* ... Alarm .. *pik-pik* .. *pik* ...

Range: Resident with local movements during winter. Entire Southwestern India up to 2100 m.

Habitat: Edges of evergreen forests, secondary and deciduous forests, monocultures, scrub and thickets and urban gardens.

1123. White-eared Bulbul *Pycnonotus leucotis*

Identification: 20 cm. A brown and white rather crestless bulbul with a black head and white cheek-patch. Blackish tail tipped white and yellow vent diagnostic.

Call: Much like other common bulbuls. Rendered as .. *tea-for-two* .. and .. *take-me-with-you* ...

Range: Resident with local movements. Southwards from Gujarat up to northern Maharashtra and Mumbai.

Habitat: Open dry scrub and gardens.

1127, 1128. Redvented Bulbul *Pycnonotus cafer*

Identification: 20 cm. A blackish-brown bulbul with black slightly crested head. Scaly patterned plumage and scarlet vent. In flight white rump and tail tips diagnostic. The race *humayuni* is paler than the race *cafer* with broader pale edges to the feathers on back.

Call: A familiar .. *peek-a-boo* ... Alarm .. *pit-pit-pit* .. *pit* ...

Range: Resident. Entire Southwestern India up to 1500 m. The race *humayuni* occurs in Mumbai and further north.

Habitat: Open forests and monocultures of all types, scrub, cultivation, groves, mangrove, beaches and urban gardens.

1135. Yellowthroated Bulbul *Pycnonotus xantholaemus*

Identification: 20 cm. A plain grey-brownish uncrested bulbul with bright yellow crown, chin, throat, flanks and vent. Brown wings and tail; the latter tipped yellow. Secretive. Scrub jungle.

Call: Described as similar to whitebrowed bulbul's.

Range: Resident (?). One record from Anaimalai hills.

Habitat: Stony country with thorn and dry scrub.

1138. Whitebrowed Bulbul *Pycnonotus luteolus*

Identification: 20 cm. A plain whitish and olive uncrested bulbul with a distinctly white forehead and supercilium. Secretive. Bush dwelling.

Call: A loud .. *poik-poik-peek* .. *poik-poik-peek* .. *picka-picka-peek* .. from within cover.

Range: Resident. Entire Southwestern India up to 1200 m.

Habitat: Secondary moist thorn forests on the coast, dry scrub and thickets, dry monocultures with dense undergrowth, estuarine scrub and beaches.

1143, 1144. Yellowbrowed Bulbul *Iole indica*

Identification: 20 cm. An olive-yellow uncrested bulbul of the forest canopy. Bright yellow forehead and supercilium diagnostic. Noisy pairs or in mixed flocks. The race *ictericus* is duller and paler than the race *indicus*.

Call: Double-noted .. *preek-preek* .. and .. *pr-pr-pree-prio* ... Difficult to separate from fairy bluebird and bronzed drongo when in mixed flocks.

Range: Resident. Western Ghats from southern Maharashtra up to 2000 m. The two races are separated at Goa, *indicus* being widespread in the south.

Habitat: Evergreen forests, secondary and moist deciduous forests, monocultures of teak, rubber and arecanut, estates and occasionally in hill gardens.

1149. Black Bulbul *Hypsipetes leucocephalus*

Identification: 23 cm. A fairly large slaty-black slightly crested bulbul readily identified by its crimson beak and legs. Noisy flocks in tall forests, especially in the higher elevations. Fly high.

Call: Noisy squeaks especially in flight .. *preeky-preeky* ... Another .. *seee* .. often heard when foraging.

Range: Resident with local movements. Western Ghats from Mumbai southwards from 400 m up to 2100 m.

Habitat: Evergreen forests (including the montane forests), secondary moist forests and plantations in the wet zones.

Family MUSCICAPIDAE

A large family of small to medium-sized insectivorous birds. Further divided into 4 subfamilies based on juvenile plumage and structure of beak, legs and feet.

Subfamily TIMALIINAE: Small to medium-sized birds with strong legs and feet. Loose and fluffy plumage. Feed on ground or among low vegetation. Noisy flocks. Sexes alike.

NOTE: Sibley and Monroe (1990) include subfamilies TURDINAE and MUS-CICAPINAE (except Monarch, Fantail and Paradise Flycatchers) under family MUSCICAPIDAE. TIMALIINAE is treated as an independent subfamily along with warblers under the family SYLVIIDAE. The Fantail and Wren Warblers are in a third family, viz. CISTICOLIDAE.

1154, 1155. Puffthroated Babbler *Pellorneum ruficeps* (Spotted Babbler)

Identification: 15 cm. A small secretive terrestrial babbler more often heard than seen. Brown above with reddish cap. White supercilium. White throat. Buffy underparts streaked and dotted with brown. Pinkish legs. Pairs. Running about quietly under thick cover. **Immature**: Duller. Lacks spots on the underparts. The races differ chiefly in the colour of crown which is richer chestnut (vs. rufous) in the race *olivaceum* than in the race *ruficeps*.
Call: A rather monotonous .. *he'll beat you* .. repeated from the same spot, similar to the common iora's. A musical song .. *ti-di-di ... ti-di-di ... ti-di-di .. tee ...* Also harsh alarm notes rendered as .. *chr-r ...*
Range: Resident. Entire Southwestern India up to 1800 m. The race *olivaceum* occurs south of the Palghat gap.
Habitat: Evergreen forests, secondary open forests with thickets, deciduous forests and bamboo.

1173, 1174. Indian Scimitar Babbler *Pomatorhinus horsfieldii*

Identification: 22 cm. A medium-sized chocolate-brown and white babbler. Longish curved yellow beak, white supercilium, throat and breast diagnostic. Pairs or noisy flocks on forest floors or low vegetation. The races differ in colour, race *horsefieldii* being greyer than race *travancorensis*.
Call: Loud bubbling .. *ook-ook-ook-ook* .. in duet. Another .. *proik* .. when alarmed.
Range: Resident. Western Ghats up to over 2000 m. The two races are separated at Goa, the race *travancorensis* representing the species in the south.
Habitat: Evergreen forests, secondary and deciduous forests with bamboo even in drier environs, tea estates and cardamom plantations.

1219, 1220, 1221. Tawnybellied Babbler *Dumetia hyperythra*
(Rufousbellied / Whitethroated Babbler)

Identification: 13 cm. A small olive-brown and ochraceous babbler with reddish cap and white throat. Creamy eyes diagnostic. **Immature**: White on throat not distinct. The races differ chiefly in the colour of crown which is entirely reddish-brown in race *abuensis* as compared with race *albogularis*.

Call: Sharp .. *sweech* .. *sweech* ...

Range: Resident. Entire Southwestern India up to 1800 m. The races are separated at Poona–Mahabaleshwar (Maharashtra); race *abuensis* representing the species in the north.

Habitat: Open scrub, thorn and bamboo and occasionally in the moister tracts within monocultures of rubber with undergrowth of *Lantana* and other thorny plants.

1224, 1225. Darkfronted Babbler *Rhopocichla atriceps*
(Blackheaded Babbler)

Identification: 13 cm. A small brown and white babbler of dense forest undergrowth. Readily identified by the black head, whitish eyes and underparts. Shy flocks. The races differ chiefly in the colour of crown; being sooty-brown in *bourdilloni* vs. black in *atriceps*.

Call: Harsh, though feeble .. *chur-r* .. *chur-r* .. turning into a louder rattle .. *tre-re-re* .. *tr* .. when alarmed.

Range: Resident. Western Ghats from Goa southwards up to 1800 m. The race *bourdilloni* represents the species south of the Palghat gap.

Habitat: Evergreen and secondary moist forests with dense undergrowth of reeds, bamboo, cane, etc., and also in monocultures including eucalyptus choked by secondary undergrowth in the moister parts.

1231. Yellow-eyed Babbler *Chrysomma sinense*

Identification: 18 cm. A small rufous-brown and white babbler with prominent white supercilium and white underparts. Yellow eyes and long graduated tail diagnostic. Feeble and jerky flight reminiscent of a wren warbler. Low bushes.

Call: Rendered as a plaintive .. *cheep-cheep-cheep* ... Alarm .. *churr* ...

Range: Resident. Widespread in Southwestern India along the eastern sides except in Kerala.

Habitat: Dry scrub, bamboo and cultivation.

1254. Common Babbler *Turdoides caudatus*

Identification: 23 cm. A medium-sized earthy-brown babbler with a long graduated tail. Heavy streaking on plumage, white throat and barred tail diagnostic. Flocks on ground.
Call: Rendered as .. *which-which-whichi-ri-ri-ri-ri-ri* ...
Range: Resident. Local. Drier sides of Southwestern India. Also recorded from the coast between Honavar and Kumta in northern Karnataka.
Habitat: Dry scrub and cultivation.

1258. Large Grey Babbler *Turdoides malcolmi*

Identification: 28 cm. A fairly large greyish babbler with white outer tail feathers (in flight). Faintly mottled above and cross-barred on tail. Bright yellow eyes diagnostic. Noisy flocks in dry open areas.
Call: Harsh .. *ke-ke-ke* .. *ke* ...
Range: Resident. Widespread in Southwestern India south of Gujarat up to 1200 m.
Habitat: Dry scrub, edges of open deciduous forests and cultivation.

1259, 1260. Rufous Babbler *Turdoides subrufus*

Identification: 25 cm. A medium-sized rufous and olive-brown babbler with rusty underparts. Grey forehead and slightly cross-barred tail diagnostic. Noisy flocks in undergrowth and bushes. The races differ primarily in colour; *hyperythrus* being richer than *subrufus*.
Call: Shrill .. *tree-ree-ree* .. *tree-ree-ree* ... Also single .. *tree* .. *tree* ... Rather similar to the Wynaad laughing thrush.
Range: Endemic. Resident. Western Ghats. Mahabaleshwar (Maharashtra) southwards and extending eastwards in Tamil Nadu hills up to over 1000 m. The race *hyperythrus* is restricted to the hills south of the Palghat gap.
Habitat: Thickets bordering evergreen forests, scrub, bamboo and *Lantana* bushes within deciduous forests and monocultures.

1262, 1263, 1264. Jungle Babbler *Turdoides striatus*

Identification: 25 cm. A medium-sized plain earthy-brown babbler with creamy eyes, yellow beak and legs. Streaks on breast. Noisy flocks in open forests. The races differ in overall coloration. Race *orientalis* has greyer tail contra rufous than in the races *somervillei* and *malabaricus* Race *somervillei* has rump paler than back.
Call: Harsh squeaks .. *kek* .. *kek* .. *ke-ke-ke* ...
Range: Resident. Western Ghats up to 1200 m. The races *somervillei* and *malabaricus* are separated in northern Karnataka. *Malabaricus* represents

the species in the south. Both races intergrade with race *orientalis* in the east at the Nilgiris, Palni hills and Maharashtra.

Habitat: Edges of evergreen forests, secondary open forests, deciduous forests with scrub, bamboo and thorny thickets, monocultures, cultivation and gardens.

1267. Yellowbilled Babbler *Turdoides affinis* (Whiteheaded Babbler)

Identification: 23 cm. A medium-sized pallid sandy-grey babbler with creamy crown. Fluffed appearance and flocks hopping about gardens and homesteads characteristic. Faint streaks on back, pale eyes and creamy legs and beak. Southern populations lack the contrast between crown and rest of the upperparts. Confusable with jungle babbler. Shrill call diagnostic.
Call: Shrill musical .. *trr-ri-ri-ri* .. *ri-ri-ri* .. *tirr* .. *tirr* ...
Range: Resident. Southwards from Goa up to 1000 m.
Habitat: Open forest with thickets, monocultures, groves, cultivation, marshes, beaches and urban gardens.

1287. Wynaad Laughing Thrush *Garrulax delesserti*

Identification: 23 cm. A medium-sized dark blackish-brown and chestnut babbler. White throat and grey breast. Lack of a white or pale supercilium diagnostic. Yellow beak and whitish legs and feet. Noisy. Flocks in dense undergrowth of evergreen forests. The largest and only laughing thrush of lower elevations.
Call: Noisy squeaks rather similar to rufous babbler. Louder .. *pree* .. like a toy whistle .. *peo* .. *peo* .. *peo* .. or .. *tiuo* .. *tiuo* .. *tiuo* ... Some notes resemble the common myna's.
Range: Resident. Western Ghats. Widespread (but patchily) southwards from Goa up to over 2000 m.
Habitat: Dense thickets and reed brakes bordering humid evergreen forests, cardamom plantations and sometimes close to human habitation.

1307, 1308. Rufousbreasted Laughing Thrush *Garrulax cachinnans*
(Nilgiri Laughing Thrush)

Identification: 20 cm. The only laughing thrush in the higher hills of the Nilgiris. Elegant. Olive-brown, black and rufous with a prominent white supercilium. Entirely rufous-ochraceous underparts diagnostic. Noisy flocks in undergrowth and gardens.
Call: Loud and typical of the laughing thrushes .. *pe-ko-ko* .. *ke-ke-ke* .. in duet.
Range: Resident. Endemic. Nilgiri hills (Tamil Nadu) from 1200 m up to over 2000 m.
Habitat: Dense undergrowth and thickets bordering evergreen forests and montane forests, grassland interspersed with bushes and gardens.

1309, 1310, 1311. **Greybreasted Laughing Thrush** *Garrulax jerdoni*
(Whitebreasted Laughing Thrush)

Identification: 20 cm. A medium-sized laughing thrush appearing dull and much like a babbler. Darkish olive-brown with a prominent white supercilium. Greyish chin, throat and breast diagnostic. Rest of underparts rufous. Noisy. Flocks in low vegetation. Less frequently on ground. Common in Kodaikanal and Munnar towns. The races differ in overall coloration. Race *jerdoni* has black chin vs. grey in the other two races. Race *fairbanki* has a white supercilium extending beyond eyes. Shorter greyish supercilium and whitish breast and belly identify race *meridionale*.
Call: Loud .. *pu-pu-pee* .. *pee-pee-pee* .. uttered by a flock. A duet .. *peo* .. *peo* .. *pee, peo* .. *tui* .. *tui* .. *tui* .. *tui* .. *tui* .. and another feebler .. *week* .. *week* .. *worek* ...
Range: Resident. Endemic. Western Ghats south of southwestern Karnataka (Coorg) from 1000 m up to over 2000 m. Race *jerdoni* occurs in Brahmagiri Hills (Coorg). Race *fairbanki* is widespread in the Cardamom, Kannan Devan and Palni hills in Kerala and Tamil Nadu south to the Anchakovil gap. Further south in the Ashambu hills race *meridionale* resides.
Habitat: Thickets and scrub bordering evergreen forests and in gardens.

1389, 1390. **Browncheeked Fulvetta** *Alcippe poioicephala* (Quaker Babbler)

Identification: 15 cm. A small brownish babbler with grey crown and nape and whitish underparts. In mixed flocks, ascending canopy. Forages much like a leaf warbler. The race *brucei* is paler and greyer than race *poioicephala*.
Call: A rather pleasant and lively .. *is-he-al-right* ... Alarm .. *trri* .. *tr-tr-tr* ...
Range: Resident. Western Ghats up to 2100 m. The race *poioicephala* occurs south of Goa and west of Biligirirangan hills.
Habitat: Evergreen forests, secondary and moist deciduous forests, monocultures and open forests with bamboo.

Subfamily MUSCICAPINAE: Small to medium-sized active birds of the canopy and edges. Brightly coloured. Sallying flight taking insects in the air often with a loud snap of beaks. Short legs and beak. Hair-like rictal bristles. Sexes often strikingly different. Immature have spots on plumage.

1407. **Asian Brown Flycatcher** *Muscicapa dauurica* (Brown Flycatcher)

Identification: 14 cm. A small rather plain sandy-brown and whitish flycatcher perched upright and sallying off and on for insects. White throat and ring around eyes and dark legs diagnostic. At close range fine streaks on plumage visible. Sexes alike. (*See* Brownbreasted Flycatcher, 1408)

Call: Feeble .. *see* .. *see* ... Also an insect-like .. *chik-chi-ri-ri* ...

Range: Resident with local movements. Breeds in Uttara Kannada (Karnataka), Cardamom hills (Kerala), and Ashambu hills (Tamil Nadu) up to 900 m. Winter visitor elsewhere (?) and in the northern Western Ghats.

Habitat: Clearings in evergreen forests, secondary and deciduous forests, monocultures and orchards, cultivation and urban gardens.

1408. Brownbreasted Flycatcher *Muscicapa muttui*

Identification: 13 cm. A small brown and white forest flycatcher separated from Asian brown flycatcher by a brownish band across breast, yellow legs and rufous-brown rump and tail. White throat separates from rustytailed flycatcher. Sexes alike.

Range: Winter visitor. Western Ghats from Goa southwards up to 1500 m.

Habitat: Evergreen forests and associated secondary vegetation, riverine forests in the drier environs and occasionally hill gardens.

1409. Rustytailed Flycatcher *Muscicapa ruficauda*
(Rufoustailed Flycatcher)

Identification: 14 cm. A small brown flycatcher with greyish throat and breast. White abdomen and rufous tail diagnostic. Forages within foliage much like a leaf warbler. Longer tail which is often wagged identifies the flycatcher. Occasionally small flocks on short trees along roadsides. Sexes alike. (*See* Brownbreasted Flycatcher, 1408)

Call: A shrill .. *weet-weet-weet* ...

Range: Winter visitor. Western Ghats from Uttara Kannada (Karnataka) southwards through Kerala up to 1000 m.

Habitat: Evergreen and secondary moist forests; along edges and clearings.

1411, 1412. Redbreasted flycatcher *Ficedula parva*

Identification: 13 cm. A small pale ashy-grey flycatcher with white underparts. Black tail with prominent white patches on the sides diagnostic. Flicking tail up and displaying the white bases at rest characteristic. **Male**: Orange-rufous chin and throat diagnostic. Most birds arrive with rather worn-out plumage and hence the red breast is not very striking. Male of race *albicilla* has buff belly separated from red throat by an ashy breast.

Call: Rather subdued .. *tick-tick-tick* .. flicking the tail.

Range: Winter visitor. Entire Southwestern India up to 2100 m. The race *parva* is not known south of Karnataka.

Habitat: Edges of evergreen forests (sometimes inside secondary evergreen forests), open deciduous forests and scrub, monocultures with sparse undergrowth, cultivation and urban gardens.

1421. Ultramarine Flycatcher *Ficedula superciliaris*
(Whitebrowed Blue Flycatcher)

Identification: 10 cm. A very small blue and white flycatcher identified by the prominent white supercilium and white patch on sides of tail. Glistening white throat diagnostic. **Female:** Darkish grey and white without a white supercilium. White throat diagnostic.

Call: Rendered as a .. *tick* ...

Range: Winter visitor. Western Ghats south up to Goa and Karnataka (?).

Habitat: Open forests, groves and orchards.

1427. Black-and-Rufous Flycatcher *Ficedula nigrorufa*
(Black-and-Orange Flycatcher)

Identification: 13 cm. A small orange-rufous forest flycatcher with black head and wings. **Female:** Brownish head.

Call: Subdued .. *chit-chit* .. and .. *chrrr-chit-chit* .. *chrrr-chit-chit*...

Range: Resident with local movements. Endemic. Western Ghats south from Wynaad (Kerala) and Biligirirangan hills (Karnataka) till Ashambu hills (Tamil Nadu) above 700 m up to over 2000 m.

Habitat: Evergreen forests including the montane, cardamom and coffee plantations and secondary evergreen forests with dense undergrowth of cane, bamboo and *Strobilanthes*.

1435. Whitebellied Blue Flycatcher *Cyornis pallipes*

Identification: 15 cm. A medium-sized indigo-blue flycatcher of dense forests identified by its white belly and a preference for dark forest understorey. More often heard than seen. Wagging tail while calling diagnostic. **Female:** Very different. Overall appearance rather similar to the European robin. Olive-brown above with greyish head and reddish tail. Orange-rufous throat and upper breast. Rest of underparts greyish-white.

Call: A very feeble whistle .. *woi-i-i-i* .. *wee-ee-eou* .. heard when all else is silent.

Range: Resident. Endemic. Western Ghats from Bhimashanker (southern Maharashtra) to Kanyakumari up to 1700 m.

Habitat: Evergreen forests (rarely along the edge), coffee and cardamom plantations and secondary moist deciduous forests with dense evergreen undergrowth of reeds, etc.

1440. Bluethroated Flycatcher *Cyornis rubeculoides*

Identification: 14 cm. A small blue flycatcher with orange-red breast and white belly. Blue throat separates from rather similar Tickell's blue flycatcher male. **Female:** Rather different. Olive-brown and white. Pale

eye-ring. Buff throat and rusty-yellow breast are diagnostic. Smaller size and duller throat separate from female whitebellied blue flycatcher. (*See* Rustytailed Flycatcher, 1409)
Call: Rendered as rather similar to that of Tickell's blue flycatcher.
Range: Winter visitor. From Goa southwards up to 1000 m.
Habitat: Secondary evergreen or deciduous forests with dense undergrowth and in gardens.

1442. Tickell's Blue Flycatcher *Cyornis tickelliae*

Identification: 14 cm. A small blue flycatcher with orange-red chin, throat and breast. White belly. Open forests. Perched upright flicking tail. **Female**: Duller grey-blue with the orange-red of chin, throat and breast replaced by paler rufous. Bluish tail without any white separate from redbreasted flycatcher.
Call: A musical .. *tsee-see-see-see-see-sik* .. *tik* .. *tik* .. *tik* ...
Range: Resident with local movements during winter. Entire Southwestern India up to 1500 m.
Habitat: Secondary evergreen forests, moist scrub and thickets, deciduous forests, monocultures, orchards and urban gardens.

1445. Verditer Flycatcher *Eumyias thalassina*

Identification: 15 m. A medium-sized greenish-blue flycatcher with a black spot in front of eyes. Paler patches may be present on vent. Confusable with Nilgiri flycatcher. Singly or in pairs in mixed flocks. Low elevation. **Female**: Duller.
Call: Rendered as .. *tze-ju-jui* ...
Range: Winter visitor. Entire Southwestern India up to 1000 m.
Habitat: Open forests and edges of evergreen forests (occasionally within), riverine scrub and gardens.

1446. Nilgiri Flycatcher *Eumyias albicaudata*

Identification: 15 cm. A medium-sized greenish-indigo-blue flycatcher with brighter forehead and supercilium. Confusable with the verditer flycatcher. However, white bases to tail and white on under tail-coverts diagnostic. Higher elevations. Southern hills. **Female**: Duller grey-brown with a greenish tinge. White bases to tail diagnostic. (*See* Verditer Flycatcher, 1445)
Call: Musical .. *piu* .. *piu* .. *piu-pee-pee-peek* ... Rather similar to the pied bush-chat.
Range: Resident. Endemic. Western Ghats. Southwestern Karnataka (Bababudan hills) through Kerala and Tamil Nadu from 600m to over 2000 m.

Habitat: Edges of evergreen and montane forests, pine and wattle monocultures, tea and coffee plantations and within hill gardens.

1448, 1449. Greyheaded Canary Flycatcher *Culicicapa ceylonensis*
(Greyheaded Flycatcher)

Identification: 9 cm. A very small olive-green and yellow flycatcher with an ashy-grey head. Brownish wings and tail. Sexes alike. The race *ceylonensis* has darker underparts than the race *calochrysea*.
Call: Pleasant and lively .. *wee-wee-wee-wit* .. and .. *ti-ti-ti-ti-ti-ti* ...
Range: Western Ghats from about 300 m up to over 2000 m. The race *calochrysea* is a winter visitor to Gujarat and Maharashtra while the race *ceylonensis* is resident southwards from northern Karnataka.
Habitat: Edges of evergreen and montane forests (sometimes within), secondary deciduous forests with dense undergrowth, coffee plantations and hill gardens.

1451, 1452. Whitebrowed Fantail *Rhipidura aureola*
(Whitebrowed Fantail Flycatcher)

Identification: 17 cm. A medium-sized dark brown and white restless bird with fan-shaped white and blackish tail held up, dancing about among low vegetation. Conspicuous white forehead and supercilium, white breast and belly and black throat diagnostic. Sexes alike. The races are separated only in hand. The race *aureola* has outer tail feathers almost entirely white and only one central pair without white tips (vs. two in race *compressirostris*).
Call: Harsh .. *chuck* .. *chuck* ... Song musical and rendered as .. *chee-chee-cheweechee-vi*...
Range: Resident. Entire Southwestern India up to 1500 m. The races are separated in southwestern Maharashtra, race *compressirostris* representing the species in the south.
Habitat: Dry forests, bamboo, scrub, orchards and gardens.

1458. Whitethroated Fantail *Rhipidura albicollis*
(Whitespotted Fantail Flycatcher)

Identification: 17 cm. A medium-sized fantailed flycatcher rather similar to the whitebrowed fantail. Shorter and narrower white supercilium, a dark slaty band across breast, white throat, dirty white belly and whitish tips to tail diagnostic. Sexes alike.
Call: Noisy .. *which* .. *which* .. *whichee* .. *whichee* ... Confusable with little spiderhunter. Song similar to that of whitebrowed fantail.
Range: Resident with local movements. Entire Southwestern India up to 2000 m.
Habitat: Moist scrub and thickets along streams and tea plantations.

1460, 1461. Asian Paradise Flycatcher *Terpsiphone paradisi*
(Paradise Flycatcher)

Identification: 20 cm (+ 30 cm tail ribbons). A fairly large-sized black-crested flycatcher appearing much like a bulbul with long ribbons on tail. **Male**: Readily identified by the white plumage and ribbons. **Female** and **Immature**: Rufous-red back and tail with whitish belly. Lack of ribbons separates the female. The rufous male in race *leucogaster* has a paler back than in race *paradisi.*
Call: A harsh nasal .. *wheech* ...
Range: Entire Southwestern India up to 2000 m. Race *paradisi* is resident locally over the range while race *leucogaster* is a winter visitor.
Habitat: Evergreen forests, secondary and deciduous forests, scrub, monocultures and orchards, cultivation and urban gardens.

1465. Blacknaped Monarch *Hypothymis azurea*
(Blacknaped Blue Flycatcher)

Identification: 16 cm. A medium-sized blue and white flycatcher with a slightly fanned tail. Black on nape and across throat diagnostic. **Female**: Bluish head. Rest of upperparts greyish-brown without black nape and throat band.
Call: Harsh double-noted .. *sweech-which* ... Another series .. *pit-pit-pit-pit-pit,* .. *pit-pit-pit-pit-pit* ...
Range: Resident with local movements. Western Ghats up to 1500 m.
Habitat: Evergreen forests, clearings and secondary forests with scrub, deciduous forests, monocultures, coffee plantations and gardens.

Subfamily SYLVIINAE: Very small to medium-sized insectivorous birds with rather dull plumage. Often difficult to identify in the field. Small size with streaked upper plumage and short fantail. Skulking in grass: *Cisticola* and *Locustella.* Small with long graduated tail cocked up or loosely hanging: *Orthotomus* and *Prinia.* Small. Unstreaked upper plumage. Broad graduated tail often fanned out. In tall grass: *Schoenicola.* Small to medium size. Dull brown. Skulking within bush and reeds. Damp habitats. Winter: *Phragmaticola* and *Acrocephalus.* Small. Dull brown-grey. Whitish underparts. In bush or trees in dry habitats. Winter: *Hippolais* and *Sylvia.* Small. Greenish-olive upperparts. Yellow-white underparts. Prominent pale supercilium and wingbars. In canopy of tall trees. Flocks. Characteristic calls. Winter: *Phylloscopus.*

1496. Goldenheaded Cisticola *Cisticola exilis* (Redheaded Fantail Warbler)

Identification: 10 cm. A very small brown and white warbler with boldly streaked upperparts. Rufous-brown collar across nape and longer buff-

tipped tail separate from similar zitting cisticola. Lark-like flight display.
Grassy hills. **Breeding Male**: Readily distinguished from zitting cisticola by unstreaked rufous crown. (*See* zitting cisticola, 1498, 1499)
Call: Rendered as .. *scrrrrr* .. *plook* .. in flight.
Range: Resident. Western Ghats southwards from Coorg including the Biligirirangan Hills and Palnis above 900 m. Isolated records from Maharashtra.
Habitat: Grassy slopes with *Pteridium* and *Strobilanthes*.

1498, 1499. Zitting Cisticola *Cisticola juncidis* (Streaked Fantail Warbler)

Identification: 10 cm. A very small brown and white warbler with bold black
streaks on upperparts. Short fantail with bold white tips diagnostic. Characteristic display flight and call. On overhead wires beside paddyfields and
meadows. Smaller size and unstreaked underparts separate from the smaller
larks. The race *salimalii* is darker than race *cursitans* in coloration with
reddish rump and a bright rufous wash on flanks.
Call: .. *chip* .. *chip* .. *chip* .. in flight. Confusable with that of flowerpeckers.
Range: Resident. Entire Southwestern India up to 2100 m. Race *salimalii* is
restricted to Kerala where it replaces race *cursitans*.
Habitat: Tall grass, reeds, scrub in swamps, rice fields, tank margins and
cultivation.

1503, 1504. Greybreasted Prinia *Prinia hodgsonii*
(Ashy Grey/Franklin's Wren-Warbler)

Identification: 11 cm. A very small ashy-brown and white warbler foraging
in flocks among scrub and tall grass like small babblers. Loosely dangling
black and white graduated tail and greyish band (not always) across white
breast diagnostic. Smaller size, paler upperparts, almost white underparts
and whitish eyes separate from ashy prinia. The race *albogularis* has a
grey breast band in summer.
Call: Feeble .. *see-see-see* ... Song rich and much like the purple sunbird's ..
chiwee-chiwee-chiwee-chip-chip-chip ...
Range: Resident. Entire Southwestern India up to 1500 m (race *hodgsonii*).
The race *albogularis* represents the species south from Coorg (Karnataka)
through Kerala and Tamil Nadu.
Habitat: Open scrub with tall grass, deciduous forests with tall grass, open
grasslands at higher elevations, coastal scrub, mangrove, monocultures of
eucalyptus and urban campuses.

1506. Rufousfronted Prinia *Prinia buchanani*
(Rufousfronted Wren-Warbler)

Identification: 12 cm. A small grey-brown wren-warbler resembling the
greybreasted prinia. Graduated tail brownish with white tips clearly visible

in flight. Rufous forehead and crown, whitish underparts washed with red-
dish on flanks and yellow-brown eyes diagnostic.
Call: Rendered as a ... *chirrup* ... Song described as a 'trill' followed by ..
sirriget-sirriget-sirriget ...
Range: Resident with local movements. South through Gujarat to northern
Maharashtra.
Habitat: Open dry stony country with scrub and grass.

1511, 1513. Plain Prinia *Prinia inornata* (Plain Wren-Warbler)

Identification: 13 cm. A small plain sandy-brown and whitish wren-warbler
with long graduated dark tail. In flight loosely dangling tail with paler tips
to outer feathers diagnostic. Plumage colour varies between summer and
winter. The race *franklinii* is darker and browner than the race *inornata*.
Call: A feeble .. *tee-tee-tee* ... Others very insect-like .. *tlick-tlick-tlick* and ..
crik-crik-crik ...
Range: Resident. Entire Southwestern India up to 1800 m. The race *franklinii*
represents the species south from about Coorg (Karnataka) through the
Nilgiris, all Tamil Nadu and Kerala.
Habitat: Open scrub, cultivation, grass, salt marshes, urban wastelands and
campuses overgrown with low weeds.

1517. Ashy Prinia *Prinia socialis* (Ashy Wren-Warbler)

Identification: 13 cm. A small dark grey and white wren-warbler readily
identified by the blackish head, long graduated brown tail, rufous flanks
and red eyes. Tail cocked and loosely wagged from side to side as the
bird forages in low vegetation or ground. Single. (*See* Greybreasted Prinia
1503, 1504)
Call: .. *teee-teee-ti-ti-ti* .. *ti* .. and .. *chi-chi-chi* .. *chi* ... Another .. *chewing-
chewing-* .. *chewing* ...
Range: Resident. Entire Southwestern India up to 2100 m.
Habitat: Urban compounds, damp scrub and grass bordering tanks, mangrove
and estuarine sedges, beaches and cultivation.

1521. Jungle Prinia *Prinia sylvatica* (Jungle Wren-Warbler)

Identification: 15 cm. A largish long-tailed brown and whitish wren-warbler.
Almost fully white outer tail feathers and entirely cream-coloured under-
parts diagnostic. White tips to tail feathers absent in winter.
Call: Rendered as .. *pit-pretty* .. *pit-pretty* .. *pit* ...
Range: Resident. Entire Southwestern India up to 1500 m.
Habitat: Open scrub and grass and hedgerows bordering cultivation.

1535. Common Tailorbird *Orthotomus sutorius*

Identification: 13 cm. A small olive-green and white wren-warbler-like bird with cocked reddish tail. Reddish cap and a blackish collar on white throat (while calling) diagnostic. Common in gardens. **Male**: Longer central pair of tail feathers. **Immature**: Lacks the reddish cap (fledgeling).
Call: .. *Pitee-pitee-pitee* .. *pitee* .. and .. *tiun-tiun-tiun* ... Alarm .. *pit-pit-pit* .. *pit* ...
Range: Resident. Entire Southwestern India up to 2000 m.
Habitat: Urban gardens, campuses, scrub, cultivation, monocultures, deciduous forests, open thorn forests, secondary evergreen forests and sometimes in clearings within evergreen forests far away from human habitation.

1543. Pallas's Warbler *Locustella certhiola* (Pallas's Grasshopper Warbler)

Identification: 13 cm. A streaked warbler larger than zitting cisticola with short white-tipped blackish tail. Head greyish with a pale supercilium. Back brown. Rufous rump. Head and back streaked with black. Whitish underparts faintly streaked on breast and flanks. In grass rising abruptly from under foot and diving in shortly. (*See* zitting cisticola, 1498, 1499)
Call: Rendered as a .. *chi-chirr* ...
Range: Winter visitor. Recorded in Kerala.
Habitat: Grassy swamps, reed beds and rice fields.

1545. Grasshopper Warbler *Locustella naevia*

Identification: 13 cm. An olive-brown grass warbler larger than zitting cisticola identified by the lack of white tips to short brown tail. Streaked upperparts. Lack of streaking on flanks and breast and presence of specks on throat and long under tail-coverts separate from Pallas's warbler. Secretive. Skulking in grass. (*See* Pallas's Warbler, 1543)
Call: Silent in winter. Others rendered as .. *chek-chek* .. or .. *churr-churr* ...
Range: Winter visitor. Widespread in Southwestern India up to 1800 m.
Habitat: Open grass along hillsides, rice fields, swamps and reed-beds.

1546. Broadtailed Grass Bird *Schoenicola platyura*
(Broadtailed Grass Warbler)

Identification: 18 cm. A largish plain brown and whitish grass warbler identified by its unstreaked upperparts and broad blackish graduated tail with white tips (underside). Tail faintly cross-barred. Secretive. Occasionally clambering tall grass.

Call: Feeble .. *pink* .. *pink*... Others .. *chek-chek-chek* .. reminiscent of the pied bushchat's alarm and a song described as a 'shrill and sweet trill ending in a few warbling notes and .. *chaks* .. '.

Range: Resident. Western Ghats south from Belgaum–Goa through Kerala and Tamil Nadu above 900 m (possibly also lower in Kerala) up to over 2000 m.

Habitat: Open grass and scrub on hillsides, marshes and reedy swamps of the higher elevations.

NOTE: The Broadtailed Grass Bird is often considered as endemic to the Western Ghats. There is, however, an old report of this species from Sri Lanka.

1549. Thickbilled Warbler *Acrocephalus aedon*

Identification: 20 cm. A largish plain olive-brown thick-billed warbler with whitish underparts. Absence of a pale supercilium diagnostic. At a glance confusable with clamorous warbler and brown shrike by overall colour and size. Solitary in low vegetation. (*See* Clamorous Warbler, 1550)

Call: Rendered as .. *tschuk* .. *tschuk* .. and .. *chr-r* ...

Range: Winter visitor. Southwestern India through Gujarat, southwestern Maharashtra, Karnataka and Kerala.

Habitat: Reed-beds, marshes including estuaries, tall grass, undergrowth in secondary forests, tea and coffee plantations.

1550. Clamorous Warbler *Acrocephalus stentoreus* (Great Reed Warbler)

Identification: 19 cm. A largish plain olive-brown warbler with a white supercilium. White throat and buff underparts. In low vegetation. Overall appearance like the more familiar Blyth's reed warbler. Larger size diagnostic. Lack of a black band through eyes separates from brown shrike. Solitary. (*See* Blyth's Reed Warbler, 1556)

Call: Loud .. *chur-r* .. *chur-r* .. and .. *ke* ...

Range: Winter visitor over most of Southwestern India up to 1600 m. Locally resident in Maharashtra (?) and Kerala (Vembanad Lake).

Habitat: Reed and grassy swamps and estuaries.

1556. Blyth's Reed Warbler *Acrocephalus dumetorum*

Identification: 14 cm. A small plain olive-brown and whitish reed warbler resembling the clamorous warbler in overall appearance. Prominent whitish supercilium. Greyish upperparts (vs. rufous-brown) separate from similar paddyfield warbler. In general Blyth's reed warbler is more widespread and common in winter and in a wider range of habitats. Forages among low vegetation. Best identified in hand. (*see also* Booted Warbler, 1562,

1563 and Lesser Whitethroat, 1567). In hand best identified by the yellow mouth and wing formula. See Figure 1.

Call: A harsh .. *chek* ... Alarm .. *chur-r-r* ... Song heard on arrival; rendered as .. *chek-chek-chek-che-chwee-chek-chek*... Also a variety of shrike-like warblings.

Range: Winter visitor. Entire Southwestern India up to 2100 m.

Habitat: Grassy marshes, estuaries, scrub, cultivation, urban gardens, monocultures, deciduous forests and edges and clearings in evergreen forests; practically anywhere with low bushes.

1557. Paddyfield Warbler *Acrocephalus agricola*

Identification: 13 cm. A small rufous-brown reed warbler with a little paler and brighter rump. Difficult to separate in the field from Blyth's reed warbler except by less distinct supercilium and yellowish-buff underparts. A preference for wet habitats is diagnostic (*See* Blyth's Reed Warbler, 1556). In hand greyish-yellow mouth diagnostic.

Call: Rendered as a harsh .. *chr-chuck* ...

Range: Winter visitor. Widespread in Southwestern India.

Habitat: Reed-beds, tall grass, sugarcane and rice fields.

1562, 1563. Booted Warbler *Hippolais caligata* (Booted Tree Warbler)

Identification: 12 cm. A small plain brownish and white warbler rather similar to the reed warbler. Behaviour more like leaf warblers among foliage. A characteristic fantail-like dancing while foraging diagnostic. Smaller size, slender beak, paler upperparts, shorter supercilium and more rounded tail separate from Blyth's reed warbler, and a preference for foraging on small trees and bush in dry areas separate from paddyfield warbler. Best identified in hand. In hand, length of the rudimentary primary feather diagnostic. Longer than primary coverts by 4–10 mm (vs. 3 mm in Blyth's reed warbler). See Figure 2. The races differ chiefly in length of tail, being over 50 mm longer in race *rama* than in race *caligata*. Race *rama* also prefers semi-desert areas.

Call: Rendered as a harsh .. *chuck* .. *chuck* .. or .. *churr-r* .. *churr-r* .. rather similar to lesser whitethroat.

Range: Winter visitor. Entire Southwestern India along the drier eastern foothills.

Habitat: Deciduous scrub and cultivation with acacias.

1565. Orphean Warbler *Sylvia hortensis*

Identification: 15 cm. A medium-sized grey warbler with whitish underparts and striking white eyes. Darker cap, glistening white throat and white outer tail feathers diagnostic. Singly in low trees and bushes. **Female**: Paler cap.

Brownish upperparts. Larger size separats from rather similar lesser whitethroat.

Call: Rendered as a loud and deep .. *chuck* .. *chuck* ... Alarm .. *trrr* .. and others .. *chichirichich* ... A musical whistle occasionally uttered.

Range: Winter visitor. Widespread in Southwestern India as far south as Kanyakumari (Tamil Nadu).

Habitat: Dry thickets, scrub and urban orchards.

1567. Lesser Whitethroat *Sylvia curruca*

Identification: 12 cm. A small grey-brown and white warbler similar to the female Orphean warbler. Glistening white throat, dark ears, blackish-brown tail with white outer feathers and a leaf warbler-like foraging habit diagnostic. Brownish eyes and bluish-grey beak. Singly in bushes and trees. (*See* Orphean Warbler, 1565)

Call: .. *tek-tek* .. Shriller and more insect-like than Blyth's reed warbler. Song rendered as .. *chivychirri* .. *chivychirri* ...

Range: Winter visitor. Widespread in Southwestern India as far south as Tamil Nadu along the drier east.

Habitat: Scrub and thorny bushes even within urban campuses.

1575. Eurasian Chiffchaff *Phylloscopus collybita* (Brown Chiffchaff)

Identification: 10 cm. A very small olive-brown active warbler without a pale wingbar. Pale supercilium and dull whitish underparts and a preference for feeding among low vegetation render the chiffchaff different from other leaf warblers and confusable with the reed and tree warblers. Smaller size, a tendency to flock and call diagnostic. Under-surface of wings yellow. Flicking wings while foraging, characteristic.

Call: Rendered as .. *tweet* .. or .. *wheet* ...

Range: Winter visitor. Southwestern India as far south as Uttara Kannada (Karnataka) through Gujarat and Maharashtra.

Habitat: Bushes (also near water), hedges, cultivation and gardens.

1578. Tytler's Leaf Warbler *Phylloscopus tytleri*

Identification: 10 cm. A very small olive-brown leaf warbler without a wing-bar. Yellowish supercilium, cheeks and ears whitish with dusky markings and whitish underparts streaked with yellow diagnostic. Low bushes. (*See* Sulphurbellied Warbler, 1581)

Call: Described as a squeaky single note.

Range: Winter visitor. Isolated records from Khandala (Maharashtra), Londa (Karnataka) and the Nilgiris.

Habitat: None described specifically. Possibly open forests with scrub.

1579. Tickell's Leaf Warbler *Phylloscopus affinis*

Identification: 10 cm. A very small olive-yellow warbler without a wingbar, readily identified by its bright yellow supercilium and underparts. Active in canopy. Singly or in small flocks. Common in the higher hills and sholas. (*See* Tytler's Leaf Warbler, 1578)
Call: A musical .. *pit-pit-whi-whi-whi-whi* ...
Range: Winter visitor. Widespread on the Western Ghats up to 2100 m.
Habitat: Montane and evergreen forests (canopy and edges), grasslands dotted with trees along the hillsides and well-planted urban campuses.

1581. Sulphurbellied Warbler *Phylloscopus griseolus* (Olivaceous Warbler)

Identification: 10 cm. A very small grey-brown leaf warbler without a wingbar. Yellow supercilium and eye-ring, dark line through eyes and dusky-yellow underparts diagnostic. Singly or pairs. Clinging to vertical rocks or tree trunks. (*See* Tytler's Leaf Warbler, 1578)
Call: A finchlike .. *peenk* ...
Range: Winter visitor. Western Ghats through Gujarat and Maharashtra to Uttara Kannada (Karnataka).
Habitat: Rocky hillsides with dry scrub and occasionally within evergreen forests and associated habitats.

1590. Inornate Warbler *Phylloscopus inornatus*
(Yellowbrowed Leaf Warbler)

Identification: 10 cm. A very small greenish-olive leaf warbler readily identified by at least one prominent wingbar and a pale band over crown. Yellowish supercilia give the head a characteristic banded pattern. Separated from the commoner western crowned leaf warbler by greener upperparts and yellowish underparts, and also a tendency to forage at lower heights and not in evergreen forests.
Call: Rendered as .. *tiss-yip* .. or .. *chilip* ..
Range: Winter visitor. South as far as Belgaum.
Habitat: Gardens, orchards and dry deciduous forests.

1601. Largebilled Leaf Warbler *Phylloscopus magnirostris*

Identification: 12 cm. The largest of our leaf warblers. Brownish-olive above with a pale yellowish supercilium and faint wingbar. Dark streak through eyes. Yellowish underparts. Rather similar to greenish warbler. Fleshy base of beak, larger size and call diagnostic. Single. Canopy. Evergreen forests. Best separated from greenish warbler in hand. In hand outermost primary and next of nearly equal length, identify largebilled leaf warbler. In the

162 MUSCICAPIDAE

greenish warbler the outermost primary is about half the length of the next.

Call: Rendered as .. *dir-tee* .. and .. *yaw-wee-wee* ...
Range: Winter visitor. Widespread on the Western Ghats south of Goa. Most common in Kerala.
Habitat: Evergreen and moist deciduous forests.

1602–1605. Greenish Warbler *Phylloscopus trochiloides*
(Greenish Leaf Warbler)

Identification: 10 cm. A very small dull-greenish active leaf warbler with a prominent pale supercilium and a faint wingbar. Whitish underparts. Dark streak through eyes. Very common. Singly in canopy. Call diagnostic. (*see* Largebilled Leaf Warbler, 1601). The races are difficult to separate except in hand. Race *nitidus* has upperparts brighter green and underparts yellowish. Outermost primaries extending beyond a line drawn from the tip of innermost primaries identify race *viridanus*. Races *trochiloides* and *ludlowi* have outermost and innermost primaries almost in level. Greyer upperparts identify the latter.
Call: .. *sichwee* .. or .. *chi-chi-chiwee* .. and .. *chi-chi-chichiwee-chichiwee* ...
Very noisy on arrival and before leaving.
Range: Winter visitor. Entire Southwestern India up to over 2000 m, race *viridanus* being the most widespread. Race *ludlowi* has been recorded from the Nilgiris and Wynaad (Kerala). Race *trochiloides* is a straggler to Kerala and race *nitidus* is widespread in Karnataka, Kerala and Tamil Nadu.
Habitat: Practically any habitat dotted with a few trees including evergreen forests, beaches and urban gardens.

NOTE: Race *nitidus* is held by some authorities as a full species, *Phylloscopus nitidus.*

1606. Western Crowned Warbler *Phylloscopus occipitalis*
(Larger Crowned Leaf Warbler)

Identification: 10 cm. A very small greyish-olive and yellowish leaf warbler with at least one prominent wingbar and a coronal band. Paler supercilia give the characteristic head pattern as in inornate warbler. Dark line through eyes. Cheeks yellowish. Beak appearing orange from below diagnostic. Flocks in canopy often in company of other insectivorous forest birds. (*See* Inornate Warbler, 1590)
Call: .. *chick-weep* ... unmistakable within dense quiet forests.
Range: Winter visitor. Western Ghats up to 2000 m.
Habitat: Evergreen, secondary and deciduous forests, teak and arecanut plantations, mangrove and urban groves.

Subfamily TURDINAE: Small to fairly large-sized birds with very musical calls and songs. Longish, thick beaks. Long legs. Often feed on ground or among bushes. Frequently black or dark. Sexes different. Males brighter in many species. Immature with spots on plumage.

1637, 1638. Whitebellied Shortwing *Brachypteryx major*

Identification: 15 cm. A small long-legged slaty-blue bird with a prominent pale supercilium and black forehead. White belly diagnostic. Shorter tail, longer legs and foraging on ground separate from rather similar whitebellied blue flycatcher in poor light. Sexes alike. The race *major* has rufous-brown sides of abdomen and under tail-coverts (vs. slaty-blue and white in race *albiventris*).
Call: Described as loud chattering, a high whistle and a thrush-like song.
Range: Resident. Endemic. Hills of the Western Ghats from southern Karnataka (Bababudans and Brahmagiris) to Nilgiris and south. Race *albiventris* represents the species south of the Palghat gap in Kerala, the Palni Hills and Ashambu hills from 900 m to over 2000 m.
Habitat: Evergreen and montane forests.

1643. Siberian Rubythroat *Luscinia calliope* (Rubythroat)

Identification: 15 cm. A small olive-brown and whitish ground bird with a prominent white supercilium. Lores black. Pink-scarlet chin and throat bordered on sides with black and lack of white on tail diagnostic. Singly. On ground with tail cocked. **Female**: Pinkish-white chin and throat without the black border. (*See* Whitetailed Rubythroat, 1647)
Call: Rendered as a harsh .. *ke* .. like that of jungle babbler.
Range: Winter visitor. Stray record from Mumbai.
Habitat: Hedges and scrub close to water and cultivation.

1645. Bluethroat *Luscinia svecicus*

Identification: 15 cm. A small olive-brown ground bird with a prominent white supercilium and rufous-orange and black tail. Striking orange bases and black tips to tail conspicuous in flight and when cocked diagnostic. Whitish underparts with a characteristic throat and breast pattern. Hopping about water and bushes. **Male**: Blue chin and throat with a black border along sides and across breast. A second rufous band on breast. Rufous-and-white spot on blue breast. **Female**: Blue of male replaced with white. Streaks and spots on throat and breast.
Call: Harsh .. *chek* .. similar to Blyth's Reed Warbler.
Range: Winter visitor. South through Maharashtra and northern Karnataka (Uttara Kannada) to about Bangalore.
Habitat: Swamps and scrub bordering estuaries and tanks.

1647. Whitetailed Rubythroat *Luscinia pectoralis* (Himalayan Rubythroat)

Identification: 15 cm. A small slaty-brown and white ground bird with a prominent white supercilium, white tips and bases to black tail. Solitary. On ground with tail cocked. **Male**: Ruby-red chin and throat with a jet-black breast and white bases to tail separate from rather similar Siberian rubythroat. **Female**: Greyish chin and throat and lack of white bases to tail separate from male. White tips to tail, black beak and legs (vs. greyish-brown) separate from female Siberian rubythroat.
Call: Rendered as a harsh .. *ke* .. like jungle babbler's.
Range: Winter straggler recorded from Londa (northern Karnataka).
Habitat: Scrub, cultivation and marshes.

1650, 1651. Indian Blue Robin *Luscinia brunnea* (Blue Chat)

Identification: 15 cm. A small ground bird of dense undergrowth with a short tail and longish legs. Hops about flicking wings and tail. Characteristic calls diagnostic. **Male**: Readily identified by blue upperparts and rusty-rufous underparts. White supercilium and under tail-coverts. **Female**: Olive-brown upperparts without the pale supercilium. Dirty white underparts. **Immature**: Dark brown with paler spots.
Call: Loud .. *tuck-tuck-tsee* .. Alarm .. *chr-r* ...
Range: Winter visitor. Western Ghats from northern Karnataka to Ashambu hills in the south up to at least 1000 m.
Habitat: Evergreen and montane forests (often along the edges) and secondary deciduous forests with dense reed undergrowth.

1661, 1662. Oriental Magpie Robin *Copsychus saularis*

Identification: 20 cm. A medium-sized glossy black and white bird with tail cocked. White shoulders, underparts and outer tail feathers diagnostic particularly when males display. Pairs on trees and ground. **Female and Immature**: Dull grey-black and white. The race *ceylonensis* is more glossy above than the race *saularis*.
Call: .. *swee* .. and .. *see-sick-sick* .. Alarm .. *chr-r-r* ... A lively song.
Range: Resident. Entire Southwestern India up to over 2000 m. Race *saularis* represents the species in southern Karnataka, Kerala and Tamil Nadu.
Habitat: Edges of evergreen and secondary forests, deciduous forests and plantations, cultivation, estuarine and beach vegetation and urban gardens.

1665. Whiterumped Shama *Copsychus malabaricus* (Shama)

Identification: 25 cm. A long-tailed black bird readily identified by the orange-rufous underparts, white rump and outer tail feathers. Loud songs.

Within low bushes in forest. **Female**: Duller than male with grey upperparts.

Call: Unmistakable song. Much louder and lower than any other forest thrush. Alarm .. *click* .. *click* ...

Range: Resident. Western Ghats up to 700 m avoiding the high rainfall tracts.

Habitat: Secondary semi-evergreen forests, open forests with thickets, deciduous forests with bamboo and teak plantations.

1672. Black Redstart *Phoenicurus ochruros*

Identification: 15 cm. A small solitary ground bird nervously hopping about the ground flicking its rufous and brown tail. **Male**: Blackish above; darkest on face and upper back. Rufous rump, tail, abdomen and vent diagnostic. **Female**: Plain brown with a paler eye-ring. Confusable with female pied bushchat. Rufous tail flicking constantly while foraging on ground diagnostic.

Call: Rendered as .. *tucc-tucc* .. *titititic* ... Alarm .. *ee-tick* .. or .. *ee-tick- tick* ...

Range: Winter visitor. Widespread in Southwestern India as far south as the Palni Hills up to at least 600 m.

Habitat: Open forests with scrub and cultivation, stony hillsides and urban compounds.

1697. Siberian Stonechat *Saxicola maura* (Stonechat/Collared Bushchat)

Identification: 13 cm. A small brown bird with dark streaking on upperparts. A white shoulder-patch, diagnostic. Solitary. Close to water on low bush. **Male breeding**: Black and rather similar to male pied bushchat. White collar and rufous (vs. white) abdomen diagnostic. **Non-breeding Male**: White collar. Darker upperparts streaked paler.

Call: Rendered as .. *chek-chek* .. and .. *pee-tack* ..

Range: Winter visitor. Widespread in Southwestern India as far south as southern Karnataka.

Habitat: Scrub, cultivation, harvested rice fields, tank beds and estuarine scrub.

1700–1702. Pied Bushchat *Saxicola caprata*

Identification: 13 cm. A small black bird readily identified by its short tail, white shoulders, rump and lower underparts. Wings and tail flicked constantly. In flight striking white rump diagnostic. Singly or pairs. Low exposed perches. **Female**: Brown. Paler underparts. Rusty rump and under tail-coverts. **Immature** and **non-breeding Male**: Paler spots on upperparts. The races differ subtly. Race *bicolor* has white on underparts extending to abdomen. Races *burmanica* and *nilgiriensis* differ primarily in size. Race *nilgiriensis* has a longer tail (not less than 60 mm).

Call: Musical .. *peet* .. *peet* .. *pree-oo-peet* .. Alarm ... *chek* .. *chek* ...

Range: Resident with local movements. Entire Southwestern India up to over 2300 m. The resident races *burmanica* and *nilgiriensis* are separated in the Nilgiris, the latter representing the species in the south. Race *bicolor* winters as far south as northern Karnataka.

Habitat: Edges of montane forests, open forests, plantations, scrub, cultivation, estuaries, beaches, urban wastelands and gardens.

1710. Desert Wheatear *Oenanthe deserti*

Identification: 15 cm. A small sandy-brown and creamy-white ground bird with black and white tail. Pale supercilium. Black throat. Dark brownish wings with traces of white diagnostic in flight. Shy. Nervously picks food on ground. **Female**: Greyer and browner than male. Lacks black throat.

Call: Alarm rendered as .. *cht-tt-tt* ..

Range: Winter visitor reported from Pune (Maharashtra) and Trichur (Kerala).

Habitat: Cultivation and fields in dry and semi-desert country.

1719, 1720. Indian Robin *Saxicoloides fulicata*

Identification: 16 cm. A small, long-tailed black ground bird with cocked tail. White shoulders and red under tail-coverts diagnostic. In flight, fully black back separates from male pied bushchat. Pairs on ground, buildings and low vegetation. **Female**: Dark brown with no wing-patch. Cocked long tail and reddish under tail-coverts separate from female pied bushchat. The races differ chiefly in colour of male. Glossy black (vs. brownish) back identifies race *fulicata* from race *intermedia*.

Call: .. *Seet* .. and .. *sit-see* .. Alarm .. *chur-r* .. Another shrill and musical call (song?) rather like Malabar lark's.

Range: Resident. Entire Southwestern India up to 1800 m. Race *fulicata* represents the species in and south through Karnataka.

Habitat: Open forests, scrub, cultivation, monocultures, estuaries, along forts and ruins, on beaches and in urban compounds.

1723. Bluecapped Rock Thrush *Monticola cinclorhynchus*
(Blueheaded Rock Thrush)

Identification: 17 cm. A small blue-black bird with orange-rufous underparts. Back with paler scalloping. White wing patch diagnostic both at rest and in flight. Forest floors and densely wooded habitats. **Female**: Very unlike male. Olive-brown above. White below scalloped with blackish-brown. Smaller size separates from rather similar Scaly thrush in appropriate habitat. Separated from female blue rock thrush by olive-brown and not grey upperparts, whitish under tail-coverts and a preference for denser vegetation.

Call: Rendered variously as .. *peri-peri* .. *goink-goink* .. and .. *tew-li-di* ... The last has a tailorbird's quality and is repeated from the same spot. Song shrill and rendered as .. *tra-trree-trrea-tre-prua-triti-prua-tri* .. not unlike orangeheader thrush.

Range: Winter visitor. Western Ghats up to 1000 m.

Habitat: Evergreen and secondary forests, moist deciduous forests, coffee and cardamom plantations and well-planted urban campuses.

1726. Blue Rock Thrush *Monticola solitarius*

Identification: 23 cm. A medium-sized blue-grey ground bird with spots, bars and streaks on entire plumage. Solitary on rocky ground. Rather upright stance at rest. Flicking tail. **Female**: Grey-brown and whitish. Barred underparts and whitish bar on wings (in flight) diagnostic. (*See* female Bluecapped Rock Thrush, 1723)

Range: Winter visitor. Entire Southwestern India up to 2100 m.

Habitat: Open forests with rock outcrops and grass, stone bunds along reservoirs, cliffs, rocky beaches and buildings.

1728. Malabar Whistling Thrush *Myiophonus horsfieldii*

Identification: 25 cm. A largish glossy blue-black crow-like thrush beside water in forests. Bright blue patch on forehead and shoulders diagnostic. Flight dashing with characteristic call. Sexes alike.

Call: A piercing .. *wheeeet* ... A lovely song of human quality giving it the apt name 'Whistling Schoolboy'. Most frequently heard in the mornings beside hill-streams.

Range: Resident. Western Ghats up to 2200 m.

Habitat: Evergreen forests, secondary and deciduous forests with water, hill-streams, arecanut plantations and wet groves in the vicinity of towns.

1731. Pied Thrush *Zoothera wardii* (Pied Ground Thrush)

Identification: 22 cm. A medium-sized black and white thrush with yellow beak. Prominent pale white supercilium, white and black spots on entire plumage and white outer feathers on dark black tail diagnostic. Solitary on ground or low trees. **Female**: Browner without black on throat and breast. Smaller size, yellowish beak, pale supercilium and spotted upperparts separate from rather similar Scaly thrush. Female Siberian thrush has less spotting on wings, white on tail restricted to the tips and paler underparts.

Range: Winter visitor. Recorded in Karnataka, Kerala and Tamil Nadu.

Habitat: Evergreen forests, secondary forests and well-wooded campuses.

1732. Siberian Thrush *Zoothera sibirica* (Whitebrowed Ground Thrush)

Identification: 22 cm. A medium-sized dark grey thrush with a prominent whitish supercilium. Brownish beak and white-tipped tail diagnostic. Female confusable with female pied thrush. **Female**: Olive-brown and whitish. Dark brown stripe on sides of throat. Belly white. (*See* Pied Thrush, 1731). **Immature Male**: Paler grey with white spots on underparts.
Range: Winter straggler into Mahabaleshwar (Maharashtra).

1733, 1734. Orangeheaded Thrush *Zoothera citrina*
(Whitethroated/Orangeheaded Ground Thrush)

Identification: 21 cm. A medium-sized orange and grey ground bird appearing like the Indian pitta in poor light. Longer tail and white throat with black bands on sides diagnostic. On ground flicking up leaves. **Female**: Grey on back tinged olive-brown. The races differ primarily in colour pattern of head. Race *citrina* has entire head orange. White wingbar (vs. shoulder) further identifies race *citrina* from race *cyanotus*.
Call: Alarm .. *kree* .. *kree* ... Song very rich, shama-like with a lot of mimicry.
Range: Resident with local movements. Western Ghats up to 2000 m. Race *citrina* winters irregularly in Maharashtra as far south as Ratnagiri.
Habitat: Evergreen and montane forests, secondary and deciduous forests, teak, arecanut, coffee and cardamom plantations and urban gardens.

1742. Scaly Thrush *Zoothera dauma* (Nilgiri Thrush)

Identification: 26 cm. A largish olive-brown and white thrush. Dark brown crescentic spots on white breast and flanks diagnostic. Pale patch on underwing conspicuous in flight. Singly or pairs in dense evergreen forests on floor wagging tail and flicking wings. Sexes alike.
Range: Rare. Resident. Western Ghats south from northern Karnataka through Kerala and Tamil Nadu from 600 m to over 2000 m.
Habitat: Dense evergreen and montane forests.

1748. Tickell's Thrush *Turdus unicolor*

Identification: 21 cm. A medium-sized grey-brown thrush confusable with Eurasian blackbird. Ashy-grey plumage, rufous underwing and white belly identify the male. **Female**: Separated from female Eurasian blackbird by white throat, streaks on sides of throat and breast, brownish flanks and faint pale supercilium.
Range: Winter straggler into Maharashtra.

1753, 1755, 1756. Eurasian Blackbird *Turdus merula*
(Blackcapped Blackbird)

Identification: 25 cm. A medium-sized dark slaty-grey-brown bird with bright yellow beak, eye-rim and legs. Solitary. On ground and on low trees.
Female: Duller and more ashy-grey (*see* Tickell's Thrush, 1748) The races differ in the colour of cap which is black contrasting sharply with the rest of upperparts in race *nigropileus*. Cap more or less same colour as back in races *simillimus* and *bourdilloni*, the latter being darker on the whole.
Call: .. *kree-ee* .. *kree-ee* ... Song melodious and intermediate in quality between Oriental magpie robin and whiterumped shama.
Range: Resident moving locally. Southwestern India up to over 2000 m. Race *nigropileus* represents the species northwards from Jog Falls (Karnataka), scattering in winter over the entire Western Ghats. Race *simillimus* occurs in Coorg and through southern Karnataka (Biligirirangan hills) to Nelliampathy hills (Kerala) and Palni hills. Race *bourdilloni* occupies the range south of race *simillimus* including the Ashambu Hills.
Habitat: Evergreen and montane forests, secondary and deciduous forests, cultivation, monocultures including teak and pine and hill gardens.

Family PARIDAE

Small active birds with short beaks. Wings rounded and weak. Tail shortish with white outer feathers. Foraging in pairs or flocks among foliage. Often clinging and hanging upside down. Sexes alike.

1794, 1795. Great Tit *Parus major* (Grey Tit)

Identification: 13 cm. A small blue-grey, black and white active bird readily identified by white cheeks and nape, black crown, throat, centre of breast and abdomen. In flight white outer feathers on black tail diagnostic. The races differ primarily in colour, race *mahrattarum* being darker, with a heavier beak, black central tail feathers, blackish wing-coverts, and less white on outer tail feathers than race *stupae*.
Call: A rather harsh .. *che-che-chi-chi* ... Others .. *weechichi* .. and .. *titiwisee* ...
Range: Resident. Entire Southwestern India up to over 2000 m. Race *mahrattarum* represents the species in Kerala.
Habitat: Edges of montane forests, deciduous forests and associated monocultures including teak and eucalyptus, orchards, scrub and urban gardens.

1798. Whitenaped Tit *Parus nuchalis* (Whitewinged Black Tit)

Identification: 13 cm. A small and unmistakable glossy black and white tit readily distinguished from great tit by the black back and yellowish tinge on underparts (not always). White nape and black band from throat

downwards separate from other smaller black and white birds, especially barwinged flycatcher-shrike. Semi-desert.
Call: Rendered as .. *tee-whi-whi-whi* ... Another described as .. *whew-whew-whew-whew* ...
Range: Resident. Local. Biligirirangan hills (Karnataka and Tamil Nadu).
Habitat: Dry scrub with acacia.

1810, 1811. Blacklored Tit *Parus xanthogenys* (Yellowcheeked Tit)

Identification: 14 cm. A small active black, olive and yellow tit with a prominent black crest. Yellow supercilium, cheeks and spots on wings. Blackish band down yellow underparts diagnostic. The races *aplonotus* and *travancorensis* differ chiefly in colour. Race *travancorensis* is greenish above. **Female**: variable. Crown and ventral band can be olive or black.
Call: .. *Chee-chee-chik* .. Song rendered as .. *cheewit-pretty-cheewit* ...
Range: Resident. Western Ghats from 500 m up to 1500 m. Race *aplonotus* represents the species north of Mahabaleshwar in Maharashtra and Gujarat.
Habitat: Secondary and moist deciduous forests, edges of evergreen forests, teak and coffee plantations and gardens in hilly towns.

Family SITTIDAE

Small active birds with short tails. Clambering and running about tree trunks and branches in forested habitats. Longish sharp beaks and long wings. Bluish or grey upperparts. Sexes alike.

1830. Chestnutbellied Nuthatch *Sitta castanea*

Identification: 12 cm. A small bluish-grey bird with chestnut underparts clambering up and down the trunks and branches of trees. Black band through eyes, white chin and white spots on tail diagnostic.
Call: .. *chip-chip* ... Song rendered as .. *wheewheewheewhee* ...
Range: Resident. Occurs locally in Maharashtra, northern Karnataka (Shimoga), Nilgiris (Mudumalai and Bandipur) and Kerala (Wynaad and Palghat) up to 1000 m.
Habitat: Open deciduous forests and on trees around villages.

1838. Velvetfronted Nuthatch *Sitta frontalis*

Identification: 10 cm. A very small purple-blue nuthatch with pinkish underparts. Crimson beak and velvety black forehead diagnostic. **Female**: Lacks the black supercilium in male.
Call: .. *chi-chi-chi-chi-chi* .. *chi* .. *chi-chichi* .. *chichi* ...
Range: Resident. Western Ghats up to over 2000 m.

Habitat: Evergreen and montane forests, secondary and moist deciduous forests, plantations and neighbourhood of hill settlements.

Family MOTACILLIDAE

Medium-sized birds with long legs and tail. Mostly ground feeding. Sexes alike. **Pipits**: Brownish lark-like birds with streaks on plumage. Tail shortish. Flocks. **Wagtails**: Grey and yellow to black and white long-tailed birds. Wagging tail while foraging. Solitary, pairs or flocks. Often beside water.

1852. Olivebacked Pipit *Anthus hodgsoni* (Indian Tree Pipit)

Identification: 15 cm. A medium-sized lark-like bird on ground in open forests. Ascends trees on approach. Greenish-olive upperparts with very fine streaks diagnostic and separate from tree pipit. Whitish supercilium, double wingbar and outer tail feathers. Buff underparts boldly streaked with brown. Small flocks. (*See* Nilgiri Pipit, 1870)
Call: Rendered as .. *tseep* ...
Range: Winter visitor. Entire Southwestern India.
Habitat: Open moist forests and plantations.

1854, 1855. Tree Pipit *Anthus trivialis* (European Tree Pipit)

Identification: 15 cm. A small tree pipit rather similar to olivebacked pipit. Generally separated from the latter by its preference for low elevation dry country. Browner upperparts boldly streaked and finely streaked underparts diagnostic (*see* Olivebacked Pipit, 1852). The races are separated by colour and shape of beak; olive-brown (vs. brown) upperparts and narrower beak identify race *trivialis* from race *haringtoni*.
Call: Rendered as .. *tseep* ...
Range: Winter visitor. Southwestern India as far south as Nilgiris.
Habitat: Open deciduous forests, plantations and urban campuses.

1857, 1859, 1860. Paddyfield Pipit *Anthus novaeseelandiae*

Identification: 15–17 cm. A small to medium-sized lark-like bird in open grassy areas. Brown upperparts streaked paler and buff underparts streaked brown on breast diagnostic. Dark tail with white outer feathers conspicuous in flight (*see* Tawny Pipit, 1861 and Blyth's Pipit, 1863). The races primarily differ in size. Race *richardi* is the largest with strikingly long legs. Races *rufulus* and *malayensis* are of the same size, the latter being the darkest with heavier streaking.
Call: .. *pi-pit* ... Another rendered as .. *rreep* .. (race *richardi*).

Range: Resident with local movements. Entire Southwestern India up to 1800 m. Race *malayensis* represents the species in Kerala. Race *richardi* is a winter visitor widespread on the Western Ghats through Kerala and Tamil Nadu.

Habitat: Cultivation, tank beds, estuarine swamps, grassy hillsides and urban meadows.

1861. Tawny Pipit *Anthus campestris*

Identification: 15 cm. A small pale brown lightly streaked pipit identified by the unstreaked underparts. White outer tail feathers conspicuous in flight. Difficult to distinguish from paddyfield pipit when individuals have streaked breast. Call diagnostic. In hand, length of hind claw rarely exceeding 11 mm (vs. over 13 mm in paddyfield pipit).

Call: Silent in winter. Described as a sparrow-like .. *tseeirp* .. or .. *tsirlui* ...

Range: Winter visitor. Not south of Karnataka.

Habitat: Open stony plains, coastal laterite with sparse grass, semi-desert and dry cultivation.

1863. Blyth's Pipit *Anthus godlewski*

Identification: 15 cm. A small pipit indistinguishable from paddyfield pipit except in hand. Call diagnostic. In hand hind toe shorter than paddyfield pipit's; rarely exceeding 14 mm. White on second outer tail feather triangular and 15 mm (vs. a streak in paddyfield pipit).

Call: Described as similar to that of paddyfield pipit (race *richardi*).

Range: Winter visitor. Rare. Widespread in Southwestern India as far south as Kerala.

Habitat: Dry fields, grassy wastelands and swamps.

1868, 1869. Longbilled Pipit *Anthus similis* (Rock Pipit)

Identification: 20 cm. A large pale rufous-brown long-tailed pipit of the higher hills. Streaked upperparts and a pale supercilium. Almost unstreaked pinkish-buff underparts and whitish throat diagnostic. Larger size separate from tawny pipit. Singly or in pairs. Flushed suddenly from under the foot. Short hovering flight. The races differ chiefly in colour; race *travancoriensis* being darker both above and below, than race *similis*.

Call: Rendered as .. *plip* .. *plip* ...

Range: Resident. Locally found on the Western Ghats above 1000 m in Pune (Maharashtra), Biligirirangan and Bababudan hills (Karnataka), Nilgiri, Palni and Ashambu hills (Tamil Nadu) and through Kerala. Race *travancoriensis* is restricted to the hills south of the Palghat gap.

Habitat: Grassy hillsides and crops.

1870. Nilgiri Pipit *Anthus nilghiriensis*

Identification: 17 cm. A medium-sized olive-brown pipit with buff underparts broadly streaked with dark brown on breast, abdomen and flanks. Dark streaks on back and pale supercilium. Dark brown tail with buff outer edge. High elevation. Singly or in pairs. Ascends trees when flushed.
Range: Resident. Endemic. Hills of Nilgiris and Palnis in Tamil Nadu and through Kerala in High Range, Ponmudi and Silent Valley from 1000 m to over 2300 m.
Habitat: Hillside grasslands, open forests and coffee estates.

1874. Forest Wagtail *Dendronanthus indicus*

Identification: 17 cm. A medium-sized pipit-like bird of the woodlands and forests. Olive-brown and black upperparts with two yellowish wingbars and yellowish-white unstreaked underparts with blackish breast bands diagnostic. Whitish supercilium. White outer tail feathers.
Call: .. *click* .. *click* ...
Range: Winter visitor. Southwestern India, southwards from Mahabaleshwar (Maharashtra) up to 2100 m.
Habitat: Evergreen, secondary and deciduous forests, monocultures, plantations of coffee and well-planted urban groves.

1875, 1875a, 1876, 1878. Yellow Wagtail *Motacilla flava*

Identification: 17 cm. A medium-sized olive-grey and yellow bird with two pale wingbars and long brownish tail. White outer tail feathers visible in flight. Fully yellow underparts and less dancing while foraging separate from grey wagtail. Grey-blackish head distinguish from yellowheaded wagtail. Flocks running about wagging tail. Sometimes on wires overhead. The races are difficult to separate during winter. Best distinguished by summer plumage. Black head identifies race *melanogrisea*. Of the three grey-headed races, race *thunbergi* nearly lacks the pale supercilium. Dark ear-coverts and grey (vs. bluish) head separate race *simillima* from race *beema*. In winter traces of summer plumage may be present. Thus race *melanogrisea* often shows black on head and race *thunbergi* has dark ear-coverts.
Call: Rendered as .. *wizzie* .. or *weesp* .. *weesp* ...
Range: Winter visitor. Entire Southwestern India up to 1500 m, *thunbergi* being the most widespread.
Habitat: Moist grasslands, edges of estuaries and tanks, wet cultivation and urban sewage.

1881, 1882. Citrine Wagtail *Motacilla citreola* (Yellowheaded Wagtail)

Identification: 17 cm. A medium-sized olive-grey and yellow wagtail separated from yellow wagtail by the yellow head. Underparts fully yellow. Flocks near water. The races are impossible to separate in the field in winter. In summer plumage race *citreola* is identified by darker back and a black hindcollar between yellow head and grey back, from race *werae*.
Call: Rendered as .. *chiz-zit* ...
Range: Winter visitor. Entire Southwestern India. Records south of Karnataka are of the race *werae*.
Habitat: Marshes, grassy edges of tanks and irrigated paddy.

1884. Grey Wagtail *Motacilla caspica*

Identification: 17 cm. A medium-sized grey and yellow wagtail with a pale supercilium readily identified by its solitary and forest-frequenting habits and vigorously wagging its tail, often exhibiting the white outer feathers. Whitish underparts being yellow only closer to the vent, separate from yellow wagtail. **Male breeding:** Black throat and prominent whitish 'V' on back.
Call: .. *chi-chi-chi* .. or .. *chi-chi-chich* .. *chich-chi-chich* ... In flight .. *chi-chee*...
Range: Winter visitor. Entire Southwestern India up to over 2000 m.
Habitat: Evergreen forest (roads and paths), secondary and deciduous forests, monocultures and plantations of coffee, tea, etc., fields, meadows, hill-streams, marshes, estuaries, beaches and urban sewage.

1885, 1886. White Wagtail *Motacilla alba*

Identification: 18 cm. A medium-sized white, black and grey wagtail separated from whitebrowed wagtail by grey back, white face and white underparts with a black mark on breast. The races differ in the facial pattern. White face identifies race *dukhunensis*. Race *personata* has black sides of face; white restricted to forehead and supercilium. Confusable with whitebrowed wagtail.
Call: Rendered as .. *chi-cheep* ...
Range: Winter visitor. Entire Southwestern India up to 1500 m.
Habitat: Open fields, meadows, lake margins and streambeds.

1891. Whitebrowed Wagtail *Motacilla maderaspatensis*
(Large Pied Wagtail)

Identification: 21 cm. A fairly large black and white wagtail. White supercilium on fully black head diagnostic. Band on wings, outer tail feathers, abdomen and vent white. White supercilium and a more horizontal stance

on ground separate from Oriental magpie robin. Pairs. Near water. Often
on buildings. The only wagtail found all through the year. **Female**: Duller.
Call: .. *pi-pick* ... Song described as similar to magpie Oriental robin's.
Range: Resident. Entire Southwestern India up to 2200 m.
Habitat: Rivers, streams, lake margins, marshes and wet fields, estuaries, salt
pans, beaches, urban sewage and buildings.

Family DICAEIDAE

Very small plain-coloured active birds visiting flowers and mistletoe. Sharp
short beaks. Short tails. Sexes alike.

1892, 1894. Thickbilled Flowerpecker *Dicaeum agile*

Identification: 9 cm. A very small olive-grey and whitish active bird readily
identified by its longish wings and thick finch-like beak. Whitish terminal
band on tail and fine streaking on pale breast diagnostic. Single. Extremely
restless.
Call: A distinct .. *chik* .. *chik* .. *chik* ...
Range: Resident. Entire Southwestern India up to 1000 m.
Habitat: Open secondary and deciduous forests, monocultures and groves
bordering cultivation.

1899. Palebilled Flowerpecker *Dicaeum erythrorhynchos*
(Tickell's Flowerpecker)

Identification: 8 cm. A tiny plain olive-grey and white active bird readily
identified by its pale flesh-coloured beak. Smaller size and short pinkish
beak separate from female sunbirds. (*See* Plain Flowerpecker, 1902)
Call: .. *tik-tik-tik* ... Another insect-like .. *tireee-se-se-se* ...
Range: Resident. Entire Southwestern India up to over 2000 m.
Habitat: Open secondary and deciduous forests, monocultures, cultivation,
groves and urban gardens

1902. Plain Flowerpecker *Dicaeum concolor* (Nilgiri Flowerpecker)

Identification: 8 cm. A tiny forest flowerpecker, almost identical in all
respects except colour of beak with palebellied flowerpecker. Black beak
and dark face with a paler supercilium diagnostic.
Call: .. *tik-tik-tik* ... Also an insect-like .. *tir-e-e-e* ... Difficult to separate from
palebellied flowerpecker.
Range: Resident. Western Ghats southwards from Mahabaleshwar
(Maharashtra) up to over 2000 m.

Habitat: Evergreen and secondary moist forests, moist scrub and plantations and within human habitation.

Family NECTARINIIDAE

Small active flower-visiting birds with slender, long, curved beaks. Short or long tails. Hover like hummingbirds. **Males**: brighter with a metallic sheen on plumage.

1907. Purplerumped Sunbird *Nectarinia zeylonica*

Identification: 10 cm. A small deep-chestnut active bird with yellow breast and whitish flanks. In good light, metallic green cap and shoulder and metallic purple rump and throat diagnostic. Female duller. Pairs or small flocks (*see* Crimsonbacked Sunbird, 1909). **Female**: Olive-brown and rufous *a*bove. Yellowish below. Black beak. Grey throat separates from female purple sunbird. (*See* female Crimsonbacked Sunbird, 1909) **Call**: .. *siswee, siswee* ... Also a tailorbird-like .. *tity* .. *tity* .. *tit-tit-tit* ...
Range: Resident. Entire Southwestern India (except Gujarat) up to over 2000 m.
Habitat: Open forests, deciduous scrub, thickets, monocultures, cultivation, mangroves, beaches and urban gardens.

1909. Crimsonbacked Sunbird *Nectarinia minima* (Small Sunbird)

Identification: 8 cm. A miniature version of purplerumped sunbird on forest canopy. **Male**: Separated by smaller size, reddish-maroon back, dark band extending beyond throat to breast and almost whitish abdomen and flanks. In clear light absence of metallic green shoulder-patch diagnostic. Pairs or in mixed flocks. **Female**: Olive above and dull yellow below. Smaller size and crimson-brown rump separate from other female sunbirds.
Call: More agitated than purplerumped sunbird .. *si-chi-chiwee* .. *si-chi-chiwee* .. *si-chi-chiwee* .. Also a tailorbird-like .. *tit-tit-tit* .. *tit* .. *tit* .. *tit* ..
Range: Resident. Endemic. Western Ghats south of Gujarat up to over 2000 m.
Habitat: Evergreen and montane forests, secondary moist forests, moist thickets and scrub, moist hill plantations and within hill settlements.

1911. Longbilled Sunbird *Nectarinia lotenia* (Maroonbreasted Sunbird)

Identification: 13 cm. A medium-sized sunbird rather similar to the commoner purple sunbird. Readily identified by its noticeably long curved beak and habit of jerking head to and fro. The little spiderhunter has a longer, rather straight beak. Singly or in pairs in wooded habitats. **Male**: Very

similar and confusable with male purple sunbird. Black. Head, shoulders, back and rump glossy with a green and purple sheen. Throat metallic green turning purple on breast. In good light, maroon-crimson breast band diagnostic. Bright yellow tufts in armpits. **Non-breeding Male**: Plumage resembles female with a black band down lower plumage. Longer beak separates from purple sunbird. **Female**: Dull olive above and yellowish below. Dark blackish tail with white tips and longer beak separate from female purple sunbird. (*See* female Crimson Sunbird, 1929)
Call: .. *cheweet-cheweet-cheweet* .. and .. *which* ... Harsher than purple sunbird.
Range: Resident. Southwestern India southwards from Thane–Mumbai (Maharashtra) up to 1600 m.
Habitat: Open forests, thickets and scrub, plantations and within towns.

1917. Purple Sunbird *Nectarinia asiatica*

Identification: 10 cm. A small glossy black-purplish sunbird readily identified by its entirely glossy plumage and shortish curved beak. Yellow and red tufts in armpits. Pairs in open and drier habitats (*see* Longbilled Sunbird, 1911). **Non-breeding Male**: Olive-brown and black above. Yellowish below with a black band down middle of throat and breast. **Female**: Olive-brown above and yellow below. Full yellow underparts including throat separate from similar female purplerumped sunbird. (*See* female Longbilled Sunbird, 1911)
Call: .. *chweet* ... Also .. *which* .. *which* .. *wichee-wichee-wichee* .. *wichee*...
Range: Resident. Entire Southwestern India up to 2400 m.
Habitat: Open forests, scrub, cultivation, estuarine and beach vegetation and urban gardens.

1929. Crimson Sunbird *Aethopyga siparaja* (Yellowbacked Sunbird)

Identification: 10–15 cm (male with longer tail). A small crimson and green sunbird readily identified by the long bee-eater-like tail. Bright yellow rump, yellow streaks on crimson throat and breast and yellowish belly. Single. **Female**: Olive. Yellowish on underside. Shorter beak and short tail graduated and tipped white separate from other female sunbirds.
Call: Rendered as .. *chi-chiwee* .. similar to blacknaped monarch's.
Range: Resident with local movements. Western Ghats as far south as Uttara Kannada (northern Karnataka) and also Nilgiris.
Habitat: Open moist forests and in gardens.

1931. Little Spiderhunter *Arachnothera longirostris*

Identification: 14 cm. A largish olive and yellowish bird appearing like a large female sunbird with very long slightly curved beak. Dark tail with white tips, yellowish belly with orange on flanks diagnostic. Sexes alike. Forests. (*See* female Longbilled Sunbird, 1911)

Call: .. *chek* .. *chek* .. *chek-chek-chek-chek* .. *chek* ... Very noisy.

Range: Resident. Western Ghats southwards from Goa (Ponda) up to 2100 m.

Habitat: Evergreen and montane forests, secondary moist forests, monocultures of teak, arecanut, banana and cardamom plantations.

Family ZOSTEROPIDAE

Very small birds with short sharp beaks. Short tails. Pale greenish-yellow plumage. Flocks. Sexes alike.

1933, 1935. Oriental White-eye *Zosterops palpebrosa*

Identification: 10 cm. Small yellow-green birds in flocks among foliage. Short, slender black beak and short tail. White ring around eyes, yellow throat and vent and whitish breast and belly diagnostic. The races differ chiefly in colour and length of tail, race *nilgiriensis* being greener above and with a longer tail than race *palpebrosa*.

Call: A feeble ... *cheeen* ... A warbling song .. *chiri-chiri-chiriu* ...

Range: Resident with local movements. Southwestern India up to over 2000 m. The race *nilgiriensis* represents the species from Biligirirangan hills (Karnataka) southwards through Coorg, Nilgiris, Kerala and Palni hills.

Habitat: Evergreen and montane forests, secondary forests and plantations, cultivation, mangrove and urban gardens.

Family PLOCEIDAE

Small to medium-sized birds with short conical beaks. Males usually brighter. Flocks. Often pests on cereal crops. **Sparrows** and **Weaverbirds**: streaked plumage and a slightly divided tail; the latter known for elaborate nesting habits. **Munias**: brightly coloured with rounded or wedge-shaped tails. Many have coloured beaks.

NOTE: Sibley and Monroe (1991) treat Sparrows, Finches, Buntings, Munias, Weavers and Wagtails/Pipits under one family viz. PASSERIDAE.

1938. House Sparrow *Passer domesticus*

Identification: 15 cm. A familiar brown and black-streaked bird hopping about domestic yards. Sometimes a nuisance building nests within living quarters. Pairs or flocks. **Male:** Rufous-chestnut back and wings with black streaks. Grey crown and rump, white face, black chin, throat and breast, white shoulder patch and underparts diagnostic. **Female:** Sandy-brown and whitish. Back streaked with dark and pale brown. A pale supercilium. Absence of streaks on underparts diagnostic. (*See* Chestnutshouldered Petronia, 1949)

Call: .. *chiu-chek* ... Song .. *chir* .. *cheer* .. *chik* .. etc.

Range: Resident. Entire Southwestern India up to over 2000 m (Nilgiris).

Habitat: Urban, cultivation, forest homesteads, estuaries and beaches.

1949. Chestnutshouldered Petronia *Petronia xanthocollis*
(Yellowthroated Sparrow)

Identification: 14 cm. A slender female sparrow-like bird with whitish underparts and bluntly divided tail chirping from bare trees or overhead wires. Chestnut-rufous shoulder patch and two white bars on wings diagnostic. In good light, yellow patch on throat identifies the male. Usually solitary or in pairs. (*See* female Common Rosefinch, 2010, 2011)

Call: Jerky and lively .. *chip-chap-chip-chap* .. *chip* ...

Range: Resident with local movements. Entire Southwestern India up to 1200 m.

Habitat: Open secondary and deciduous forests, scrub, cultivation and within rural limits.

1957, 1958. Baya Weaver *Ploceus philippinus* (Common Weaverbird)

Identification: 15 cm. A medium-sized yellow, buff and brownish female sparrow-like bird streaked darker on upperparts. Yellowish supercilium. Pale streaks on yellowish breast diagnostic. Characteristic suspended nests from trees or overhead wires. Large flocks in wet fields (*see* Streaked Weaver, 1962). **Male breeding**: Yellow crown. Rest of upperparts dark brown streaked with yellow on back. Dark brown sides of face and throat and unstreaked yellow breast diagnostic. The races differ chiefly in colour, race *travancorensis* being darker above than race *philippinus*.

Call: .. *chee-ee-ee* .. and .. *chit* .. *chit* .. *chit* .. *chit* ...

Range: Resident with local movements. Entire Southwestern India up to 1000 m. Race *travancorensis* represents the species southwards from Goa.

Habitat: Cultivation, river banks, lake margins, hillsides with wild date palms, estuaries with short emergent plants, dry plains with palmyra palms and acacias and suburban wastelands.

1961. Blackbreasted Weaver *Ploceus benghalensis*
(Blackthroated Weaverbird)

Identification: 15 cm. An unmistakable weaverbird separated from all others by unstreaked whitish underparts and blackish breast. Nest like that of baya weaver but concealed within tall grass. Flocks. **Male breeding**: Yellow crown bordered with black. Rest of upperparts dark brown with paler edges. Whitish below with a dark brown breast-band. **Female** and **non-breeding Male**: Dark brown back streaked paler. Distinct yellow and buff supercilium. Dark ears and yellow patch on sides of neck separate from female common weaverbird, even if dark breast is not distinct or appears like streaks. Yellow chin and throat with a dark moustachial streak. (*See* Streaked Weaver, 1962)

Call: Rendered as .. *chit-chit* .. in flight.

Range: Resident. Local. Bhandup (near Mumbai).

Habitat: Grassy swamps and reeds.

1962. Streaked Weaver *Ploceus manyar* (Streaked Weaverbird)

Identification: 15 cm. An unmistakable weaverbird separated from the other two by almost fully streaked underparts. Yellow supercilium continued around dark ear-coverts to form a yellow collar on sides of neck, most diagnostic. Nest within tall reeds in marshes; more rounded than baya weaver and without the elongated entrance. (*See* Blackbreasted Weaver, 1961). **Male breeding**: Yellow crown. Dark brown sides of head. Rest of upperparts dark brown streaked paler. Brown throat. No yellow on underparts. Heavy streaking on whitish breast and flanks.

Call: Rendered as .. *chirt* .. *chirt* ... Others described as .. *tre* .. *tre* .. *cherrer* .. *cherrer* ...

Range: Resident. Local. Entire Southwestern India.

Habitat: Swamps, reed beds, lake and river margins with tall sedge or grass.

1964. Red Avadavat *Amandava amandava* (Red Munia)

Identification: 10 cm. A very small warbler-like brown and whitish bird readily identified by its short conical brownish-crimson beak and fleshy pink legs. Dark upperparts with a few white spots, crimson upper tail-coverts spotted with white, diagnostic. Flocks in grass and rice fields. **Male breeding**: Crimson with blackish wings, belly and tail. White spots on upperparts and tips of tail. Crimson beak. **Immature** and **non-breeding Male**: Traces of crimson and black on entire plumage. Brownish-crimson beak.

Call: A feeble .. *pink* ...

Range: Resident. Widespread in Southwestern India up to 2100 m. Occurrence in Kerala doubtful.

Habitat: Grassy swamps, rice fields, cultivation, tank beds and damp urban wastelands.

1965. Green Avadavat *Amandava formosa* (Green Munia)

Identification: 10 cm. A small warbler-like munia readily identified by its olive-green and yellow plumage and short conical crimson beak. Dark tail

and olive bars on yellow underparts diagnostic. Flocks feeding on ground.
Female: Duller with less barring on underparts.
Call: Rendered as .. *swee* .. *swee* ...
Range: Resident. Localized in Southwestern India between Gujarat and
Mahabaleshwar (Maharashtra).
Habitat: Open grassland, cultivation and scrub.

1966. Whitethroated Silverbill *Lonchura malabarica*
(Whitethroated Munia)

Identification: 10 cm. A small plain grey-brown munia with whitish underparts and pointed blackish tail. Short slaty-blue conical beak and white
rump diagnostic. Sexes alike. Flocks in dry habitats. **Immature**: Rump
spotted with brown.
Call: Feeble. Rendered as .. *chip* ...
Range: Resident with local movements. Entire Southwestern India along the
drier foothills not over 1000 m.
Habitat: Open cultivation, scrub, grasslands and urban wastelands.

1968. Whiterumped Munia *Lonchura striata* (Whitebacked Munia)

Identification: 10 cm. A small dark blackish-brown and white munia with
white rump and belly. Grey-black short conical beak diagnostic. Pale
streaks on back visible in hand. White and blackish plumage striking in
flight. Flocks. Often in forests. Sexes alike. (*See* Blackthroated Munia,
1971)
Call: .. *ptirri* .. *ptirri* ...
Range: Resident with local movements. Entire Southwestern India up to
1800 m.
Habitat: Secondary moist forests and scrub, deciduous forests, monocultures
of teak, arecanut, eucalyptus, cultivation, bamboo, lake margins, estuaries
and urban gardens.

1971. Blackthroated Munia *Lonchura kelaarti* (Rufousbellied Munia)

Identification: 10 cm. A small dark chocolate-brown munia without a white
rump diagnostic. Paler upper tail-coverts. Blackish cheeks, throat and
breast. Rest of underparts pinkish-brown. Pale streaks on back visible in
hand. Sexes alike. Flocks in hills and forest. **Immature**: Plain dark brown
above and rufous-brown below without any blackish marking on throat
and breast. Darker than non-breeding scalybreasted munia. (*See* immature
Blackheaded Munia, 1978)
Call: A nasal .. *kenk* .. rather like that of blackheaded munia and unlike that
of whiterumped and scalybreasted munias.

Range: Resident with local movements. Western Ghats southwards from Goa through Karnataka, Kerala and Tamil Nadu up to 2100 m.

Habitat: Forest and hillside cultivation, open grassland bordered by forests, tall rocky cliffs within evergreen forests and within hill settlements.

1974. Scalybreasted Munia *Lonchura punctulata* (Spotted Munia)

Identification: 10 cm. A small brown munia with slaty-black beak and white underparts speckled (scaly) with black. Face and breast dark chestnut. Olive upper tail-coverts. Whitish bars on rump. Sexes alike. Flocks in wooded habitats. **Non-breeding**: Plain brown above and sandy below without the spots. Darker beak and paler head separate from immature blackheaded munia. (*See* immature Blackthroated Munia, 1971)

Call: .. *peet* .. *peet* ...

Range: Resident. Entire Southwestern India up to 2100 m.

Habitat: Open forests and secondary scrub, plantations, grass with scrub and urban gardens.

1978. Blackheaded Munia *Lonchura malacca*

Identification: 10 cm. A small stocky munia readily identified by its rich rufous-chestnut and white plumage and black head. Underparts white with black throat, breast, centre of belly, thighs and vent. Striking conical bluish-grey beak. Sexes alike. Pairs or flocks in grass and crops. **Immature**: Rufous-brown upperparts and whitish-buff underparts. Beak and head stockier than any other munia of comparable plumage. Pale beak and head darker than back diagnostic. (*See* immature Blackthroated Munia, 1971 and non-breeding Scalybreasted Munia, 1974)

Call: A nasal .. *enk* .. rather similar to that of blackroated munia.

Range: Resident. Entire Southwestern India southwards from Maharashtra up to 2100 m.

Habitat: Cultivation (even in forest clearings), grassy swamps, estuarine marshes and urban wastelands.

Family FRINGILLIDAE

Sparrow-like medium-sized birds with thick conical beaks. Longish wings and slightly forked tails. Males brighter.

2010, 2011. Common Rosefinch *Carpodacus erythrinus*

Identification: 15 cm. A medium-sized rosy-crimson (appearing brown and rosy in winter) sparrow-like bird with brownish wings. Crimson rump. Whitish vent. Pairs or small flocks. **Female**: Like female house sparrow.

Olive-brown above; wings with two pale wingbars and whitish underparts streaked on throat and breast diagnostic (*see* Chestnutthroated Petronia, 1949). The races differ in colour of summer plumage, male of race *roseatus* being darker and purplish. Heavier streakings on underparts of female of race *roseatus*.
Call: Rendered as .. *twee-ee* ...
Range: Winter visitor. Entire Southwestern India up to over 2000 m. Race *ferghanensis* is not known south of Belgaum district (northern Karnataka).
Habitat: Open grasslands on the hills, hillside scrub and secondary forests.

Family EMBERIZIDAE

Largish sparrow-like birds with thick conical beaks. Back often streaked. Tail longish. Males brighter. Flocks in cultivation.

2043. Blackheaded Bunting *Emberiza melanocephala*

Identification: 18 cm. A big sparrow-like bird with black head, rufous back and yellow underparts. Whitish double wingbar. Large flocks in fields. **Female**: Brown above with darker streaks. Whitish-buff below. Unstreaked underparts and yellow vent diagnostic. Difficult to separate from female redheaded bunting except in hand (*see* Baya Weaver, 1957, 1958). In hand larger size and rufous-tinged (vs. yellow) rump separate from otherwise similar female redheaded bunting.
Call: Rendered as .. *tweet* ...
Range: Winter visitor. Southwestern India as far south as Karnataka.
Habitat: Open dry cultivation.

2044. Redheaded Bunting *Emberiza bruniceps*

Identification: 17 cm. A bunting slightly smaller than blackheaded bunting distinguished by its largely yellow plumage, reddish-cinnamon crown, face and throat. Black streaks on upperback and whitish double wingbar diagnostic. Large flocks in fields. **Female**: Rather sparrow-like. Yellow vent diagnostic. Separated from similar female blackheaded bunting by smaller size and yellow rump. Best differentiated in hand. (*See* Blackheaded Bunting, 2043)
Call: Rendered as .. *tweet* ...
Range: Winter visitor. Entire Southwestern India along the eastern side as far south as Tamil Nadu.
Habitat: Open dry cultivation.

2050. Greynecked Bunting *Emberiza buchanani*

Identification: 15 cm. A small sparrow-like grey, brown and rufous bird with white eye-ring and outer tail feathers. Streaked back. Brown forked tail. Yellow-orange beak. A dark moustachial streak. Underparts rufous, mottled with white. Sexes more or less alike. Flocks on ground. Pipit-like gait.
Call: Rendered as .. *click* ...
Range: Winter visitor. Drier sides of Gujarat and Maharashtra.
Habitat: Open scrub and rocky fallow land.

2060. Crested Bunting *Melophus lathami*

Identification: 15 cm. A small black and chestnut bird with a distinct bulbul-like crest and orange beak. Chestnut wings and tail. Flocks on roadsides in hills. **Female**: Like Malabar lark. Olive-brown above with darker streaks. Shorter crest. Rufous on wings and outer tail feathers. Yellowish below with dark streaks on breast. Dark moustachial streak and pale eyering diagnostic. (*See* Malabar Lark, 901)
Call: Rendered as .. *tip* ... Song described as .. *which* .. *which* .. *which-whi-whee-which* ...
Range: Resident. Local. Western Ghats in Satara (Maharashtra).
Habitat: Open stony scrub with grass and cultivation.

INDEX

PLATES

Plate 1

Plate 2

Plate 3

Plate 4

Plate 5

Plate 6

Plate 7

Plate 8

Plate 9

Plate 10

Plate 11

Plate 12 (seen from below)

Plate 13 (seen from above)

Plate 14 (seen from below)

Plate 15 (seen from below)

Plate 16

Plate 17

Plate 18

Plate 19

Plate 20

Plate 21

Plate 22

Plate 23

Plate 24

Plate 29

Plate 30

Plate 34

Plate 35

Plate 36

Plate 44

Plate 45

Plate 46

Plate 47

Plate 48

Plate 49

Plate 50

Plate 51

Plate 52

Plate 56

Plate 57 Wing Formulae of Warblers

Plate 58

Plate 59

Plate 60

Plate 61

Plate 62

Plate 63

Plate 1

50CM

Plate 2

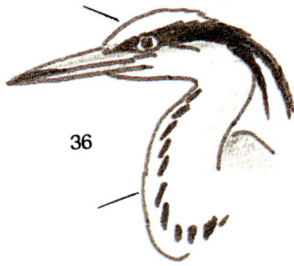

36

37

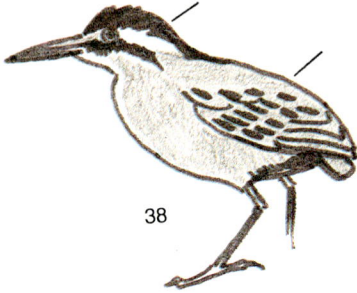

38

42

42'

52

53

56

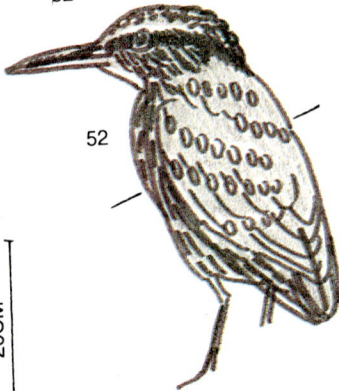

52

58

59

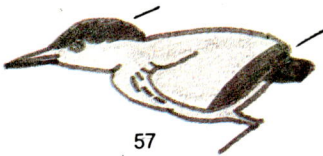

57

20CM

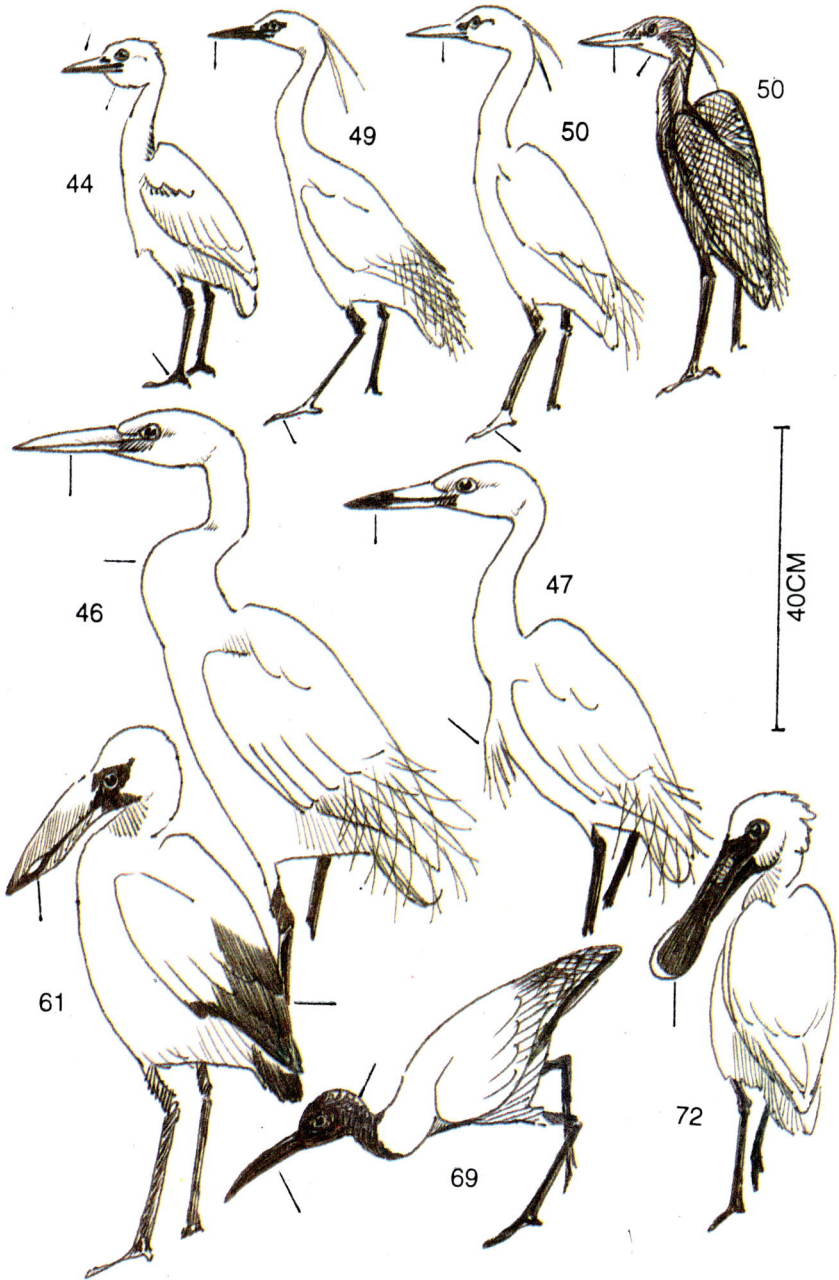

Plate 3

44

49

50

50

46

47

40CM

61

69

72

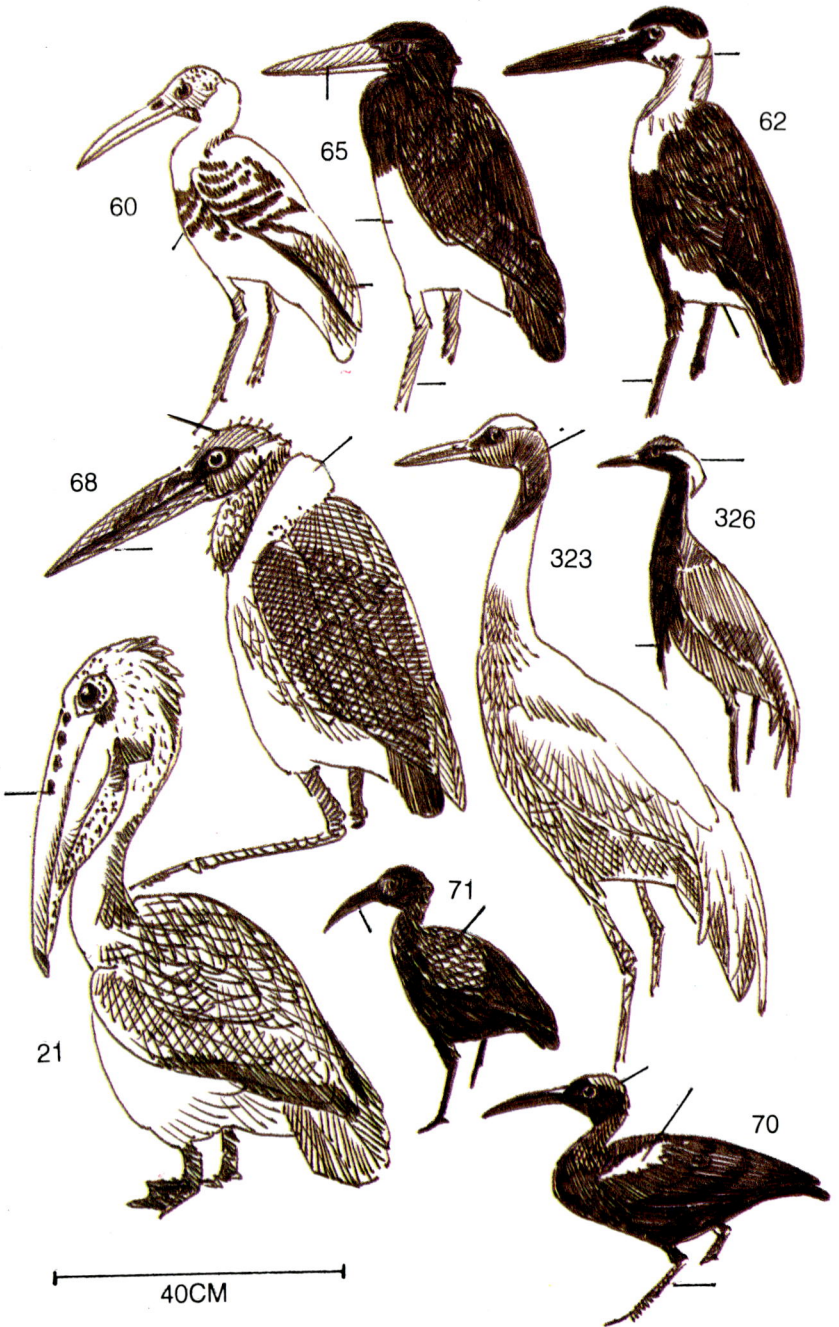

Plate 4

60

65

62

68

323

326

21

71

70

40CM

Plate 5

73

323

50CM

326

Plate 6

5

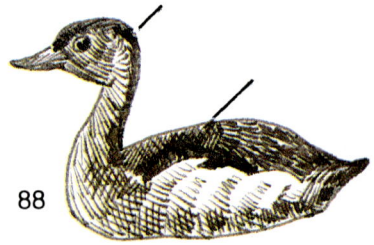

88

♀

104

♂

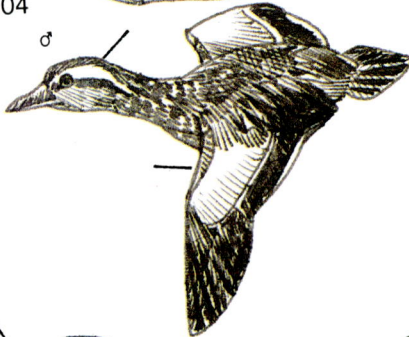

89

♀

114

350

94

♀

♂

♂

94

25CM

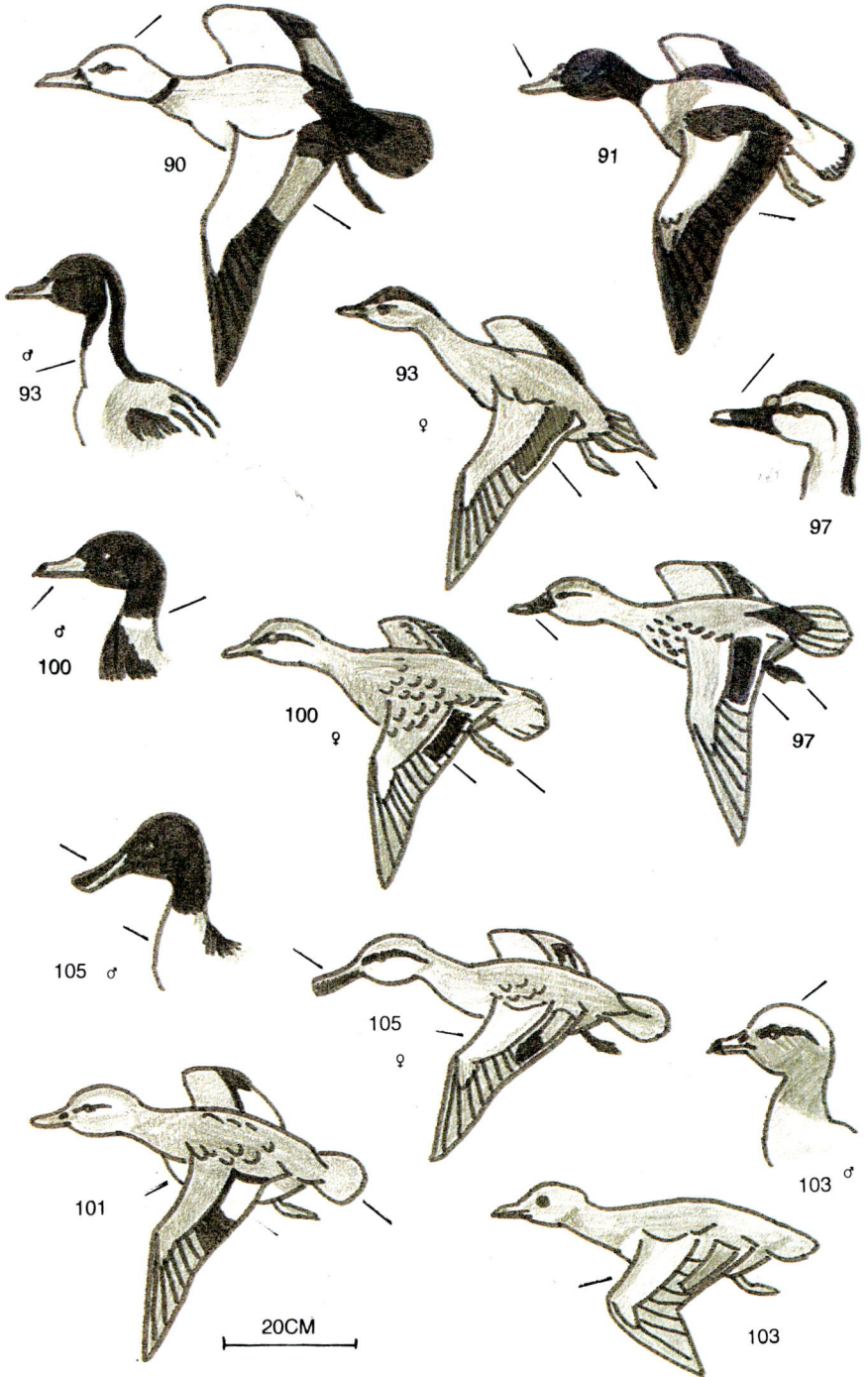

Plate 7

Plate 8

♂
107

107 ♂

107 ♀

♂
108

108 ♀

109

111
♂

112 ♂

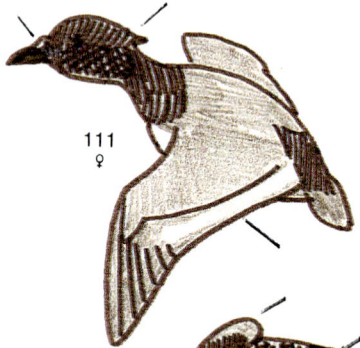

111
♀

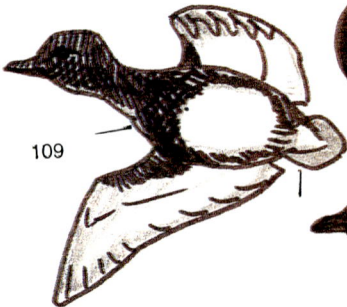

109

115 ♀

115 ♂

45CM

Plate 9

124

126

127

30CM

157

136

145

151

137

147

133

135

137 IMM

147 IMM

Plate 10

130

153

156

50CM

164

165

♂
190

190
♀

191

192

♀

193

♂

203

Plate 11

40CM

Plate 12

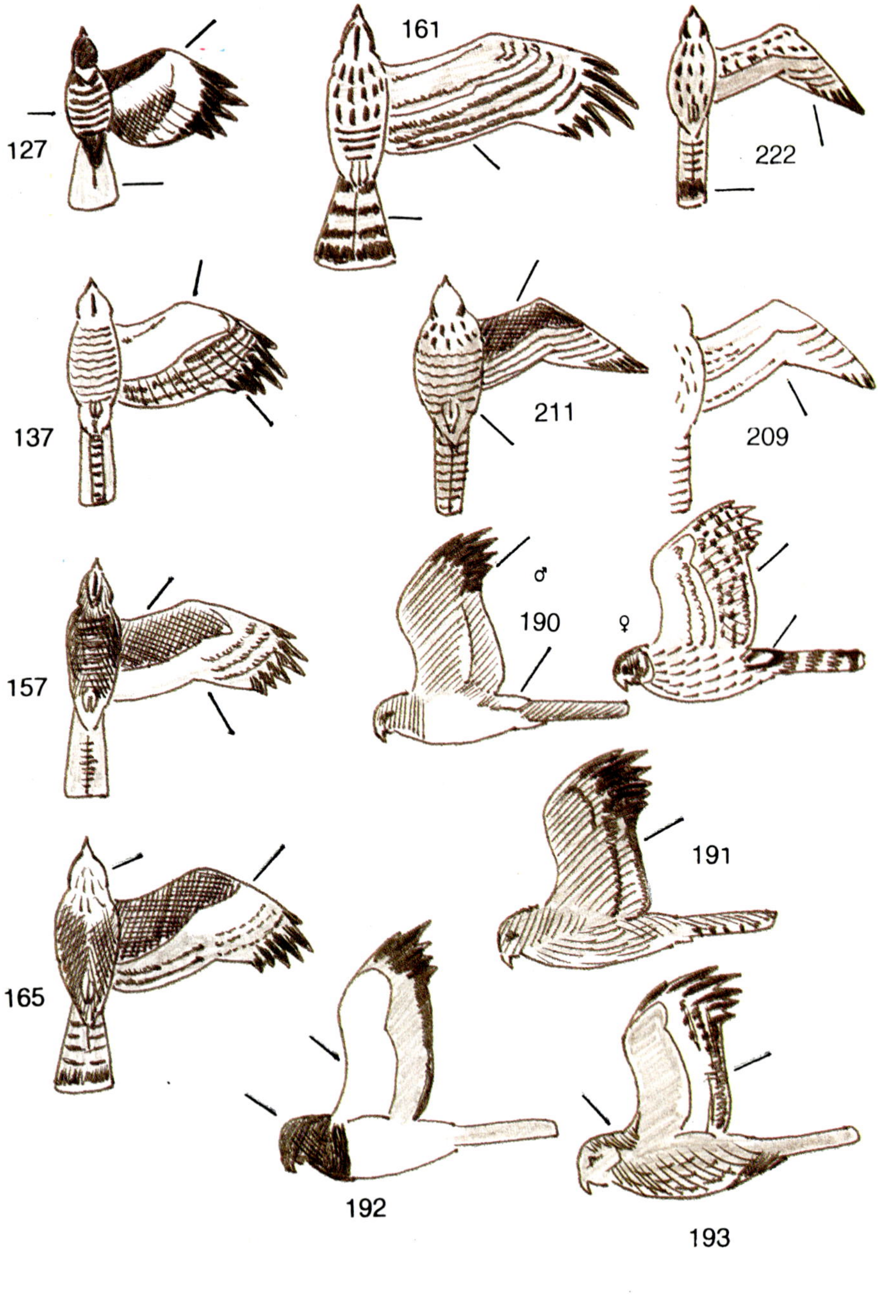

127

161

222

137

211

209

157

♂

190

♀

191

165

192

193

Plate 13

170

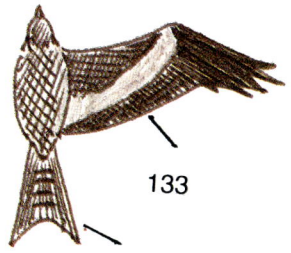

133

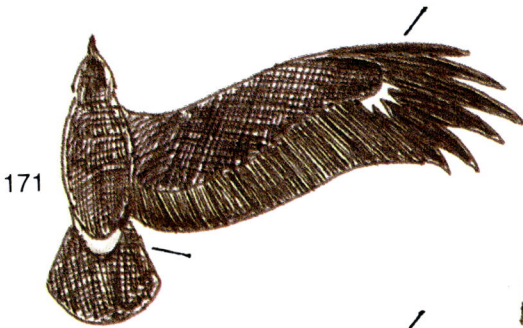

171

175

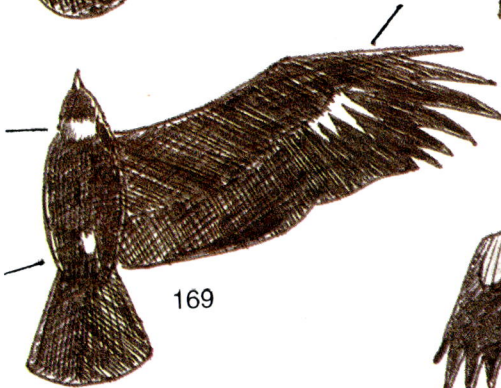

169

173

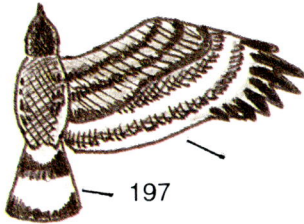

197

172

Plate 14

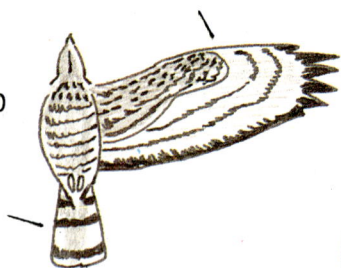

130

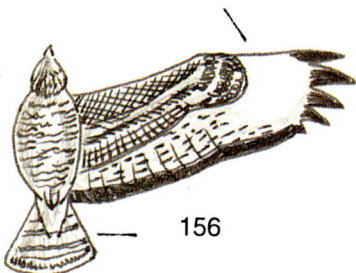

156

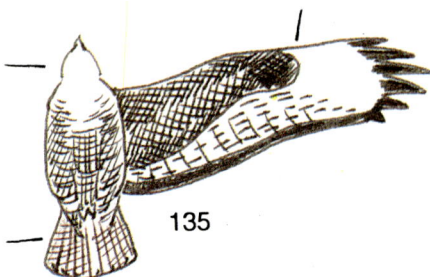

135

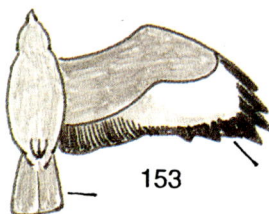

153

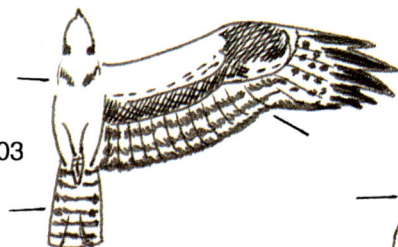

203

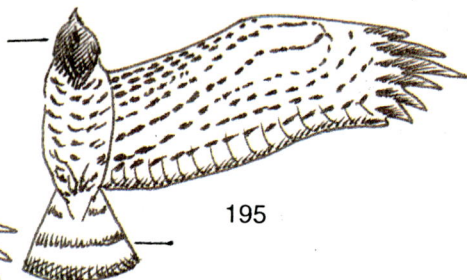

195

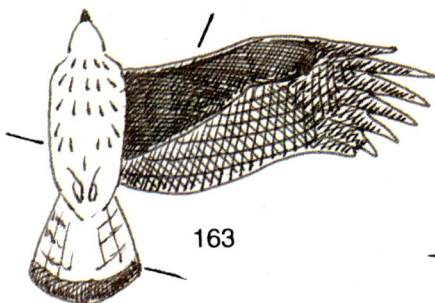

163

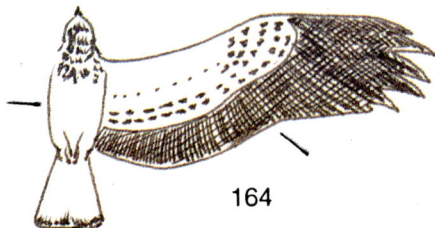

164

Plate 15

185

182

178

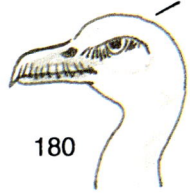

180

45CM

187

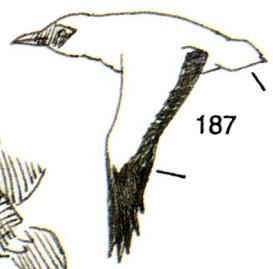

187

185

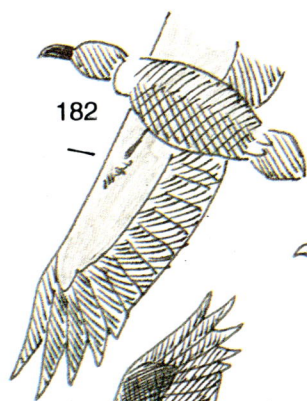

182

40CM

178

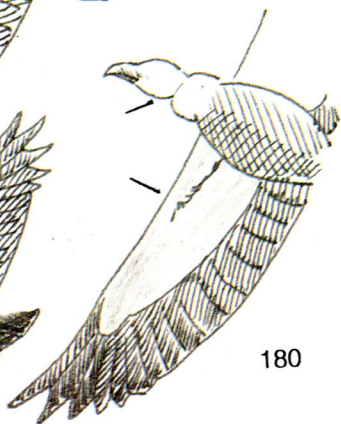

180

Plate 16

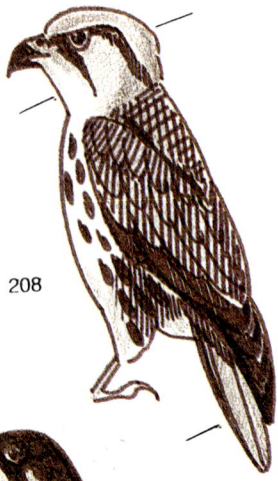

208

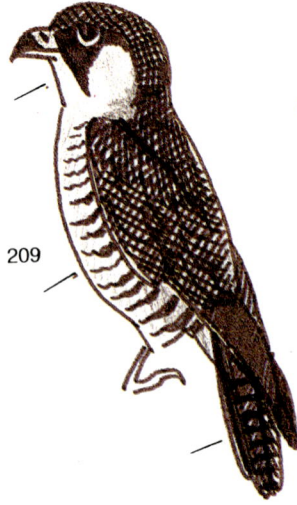

209

212

214

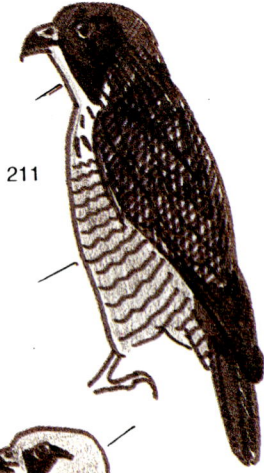

211

219

♂
220

220
♀

♂
222

♂
221

20CM

Plate 17

Plate 18

311 ♂

♀

301 ♂

301 ♀

299 ♂

299 ♀

357 ♂

357 ♀

40CM

Plate 19

313

314

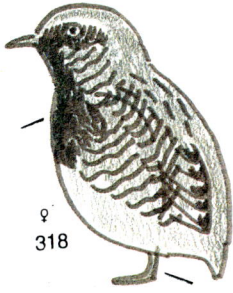

♀
318

10CM

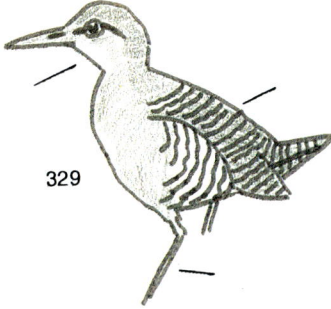

329

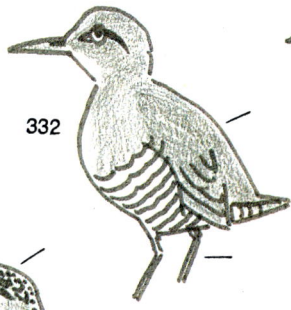

332

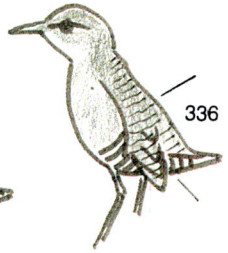

336

337

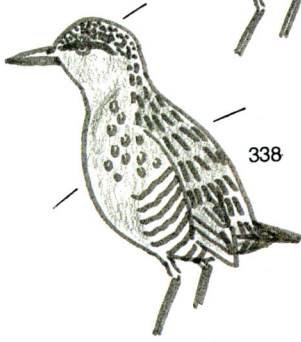

338

340

344

♀

346   ♂

347

349

25CM

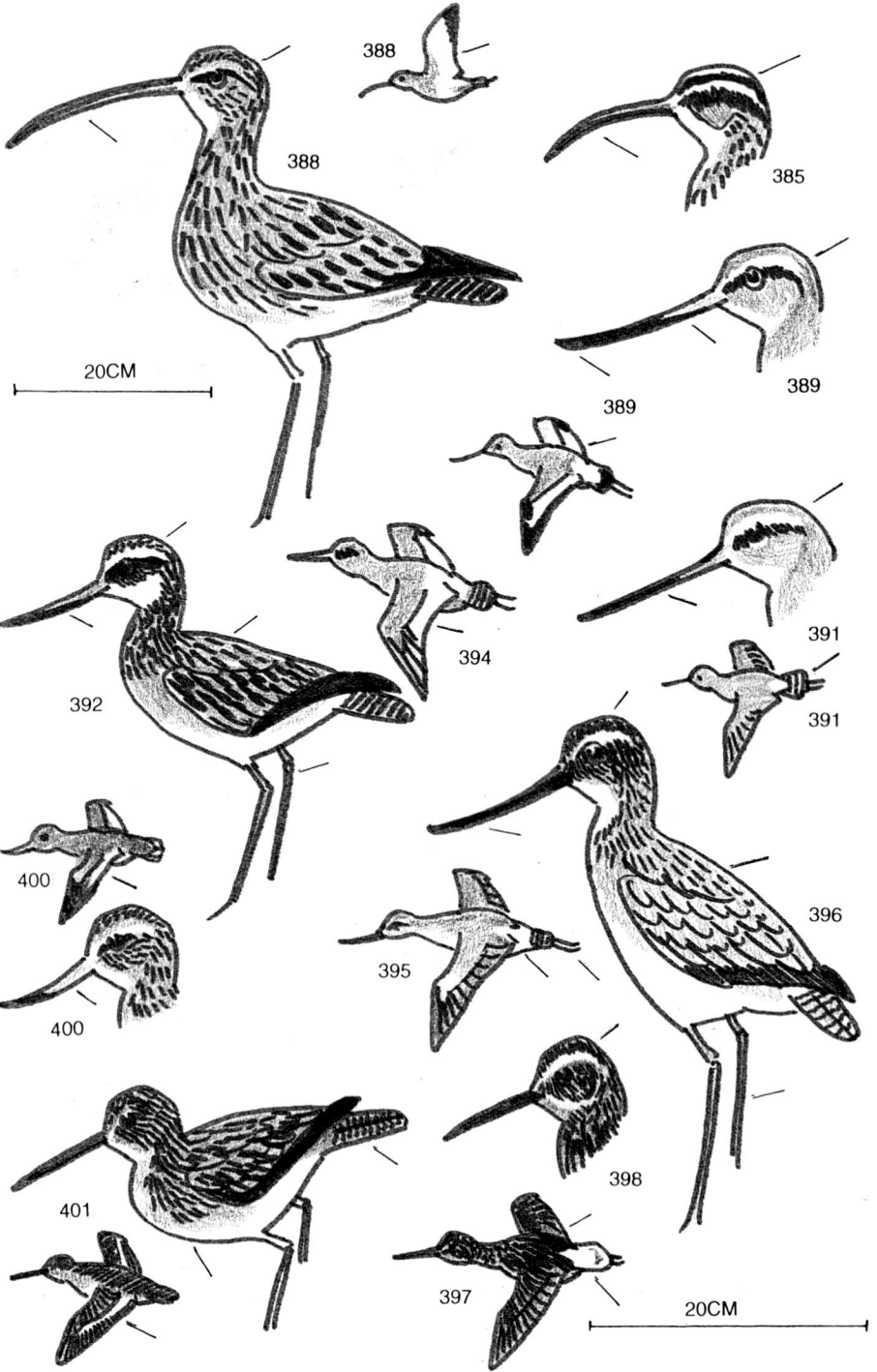

Plate 20

20CM

20CM

Plate 21

358

358

358

359

359

360

359

430

432

434

436

437

30CM

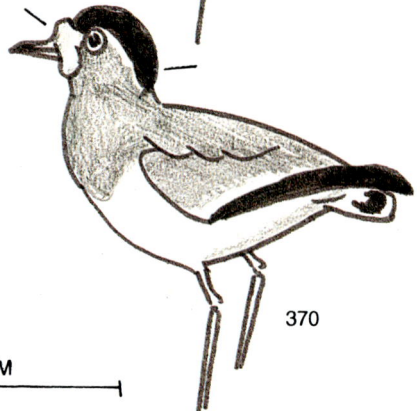

Plate 22

440

362

363

365

366

370

20CM

Plate 23

373

371

371

373

373

384

379

376

374

381

414

402

416

417

422

424

420

15CM

Plate 24

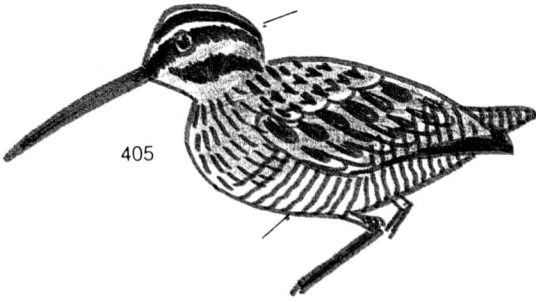

405

406

406

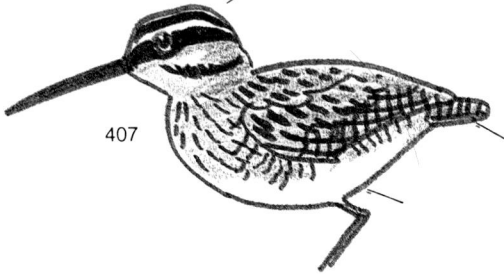

407

410

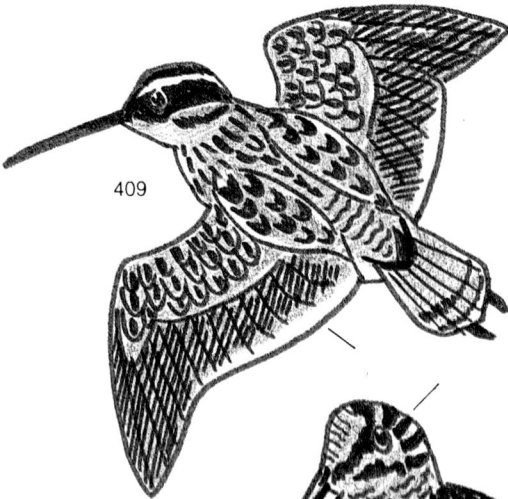

409

411

429

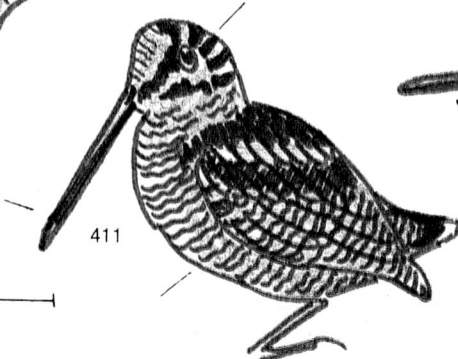

411

10CM

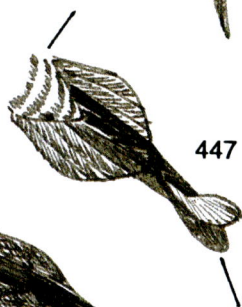

Plate 25

Plate 26

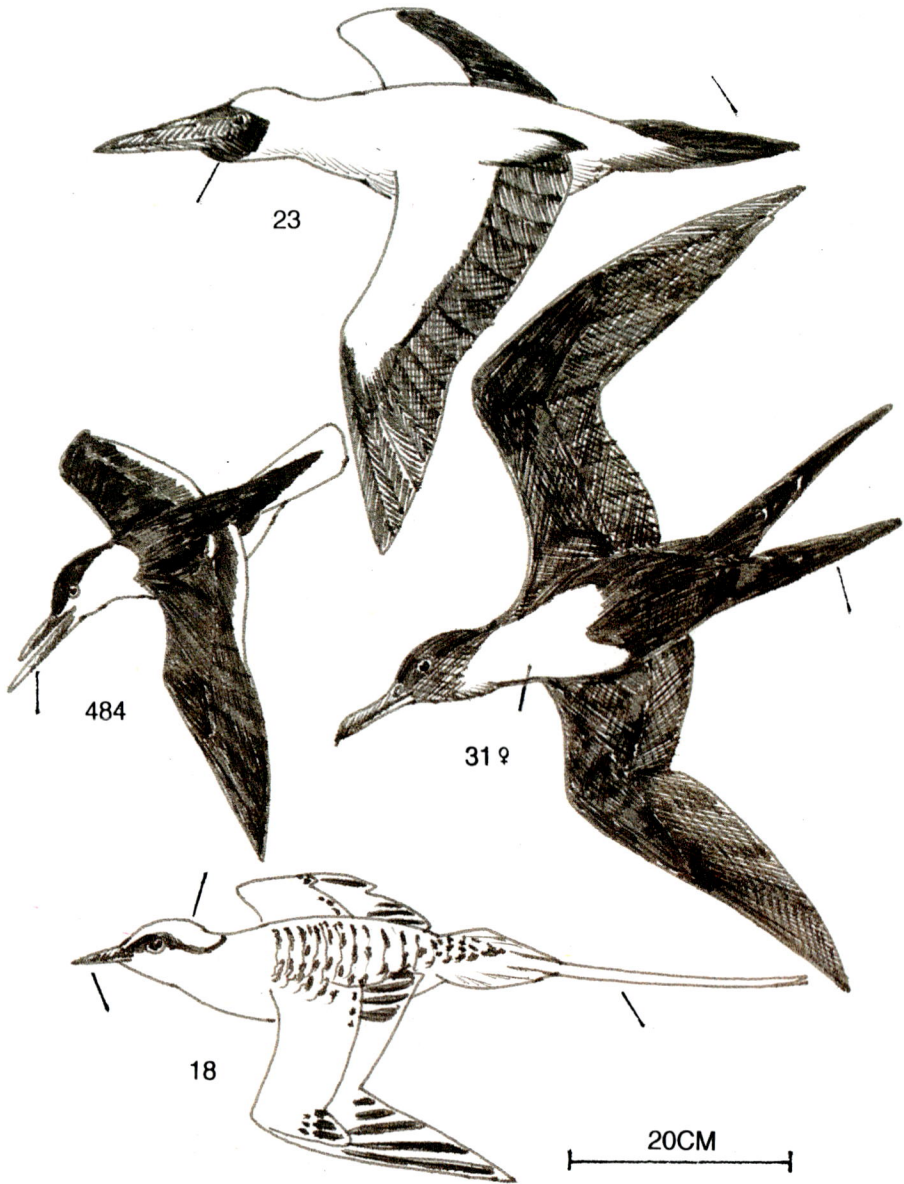

23

484

31 ♀

18

20CM

Plate 27

452

456

455

454

453

450

25CM

9

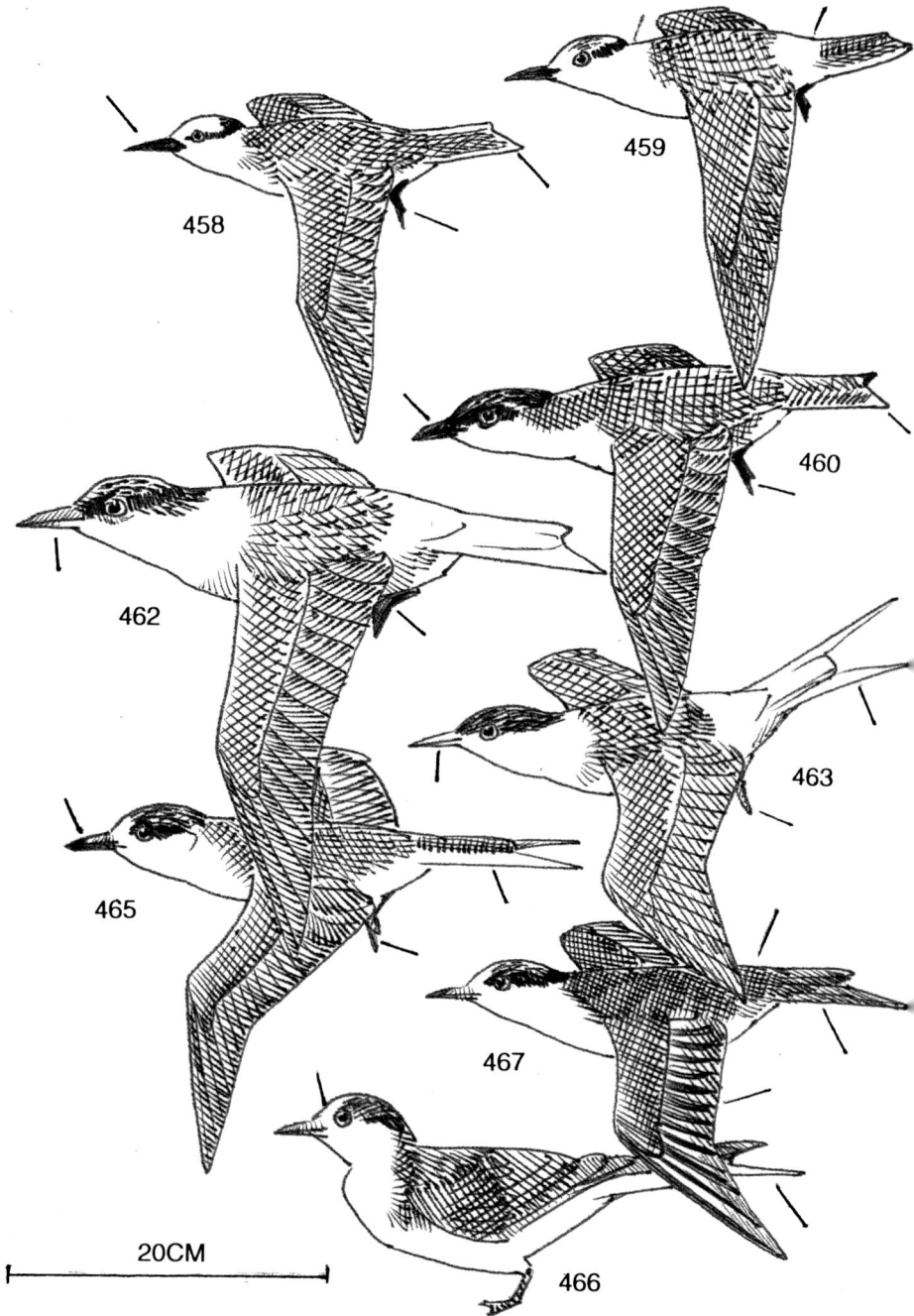

Plate 28

458

459

460

462

463

465

467

466

20CM

Plate 29

470

472

474

475

476

478

479

480

20CM

Plate 30

535 ♂
535 ♀

541

534

537

531

517

521

542

511

507

496 ♀ ♂

504

501 ♂

501 ♀

25CM

Plate 31

566

545 ♀

545
♂

550
♂
♀

20CM

558
♀

558
♂

564
♂

564
♀

595

599

564

602

605
IMM

598

Plate 32

573

572
IMM

572

573
IMM

576

♀
576

573
IMM

576
IMM

578

582

578
♀

578
IMM

584 ♀

581

♂
584

586   ♂

♀

10CM

Plate 33

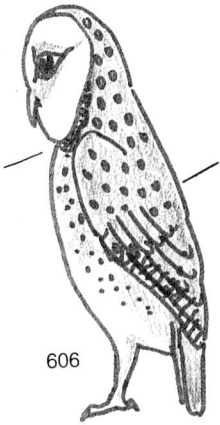

606

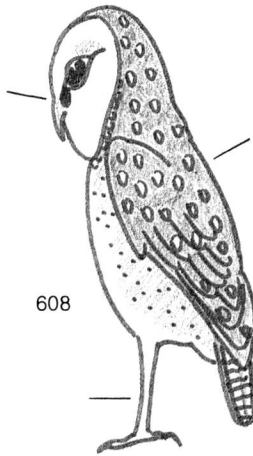

608

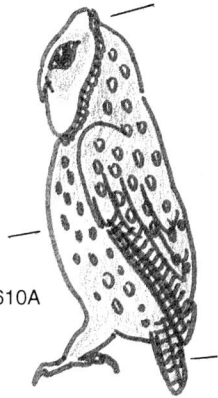

610A

614

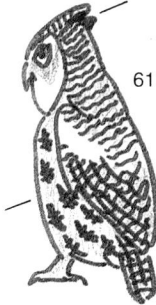

615

622

636

652

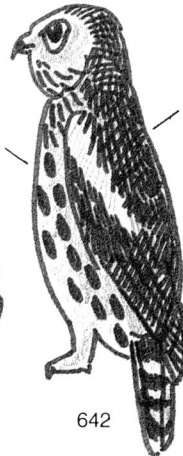

642

664

25CM

Plate 34

627

628

630

657

631

659

20CM

Plate 35

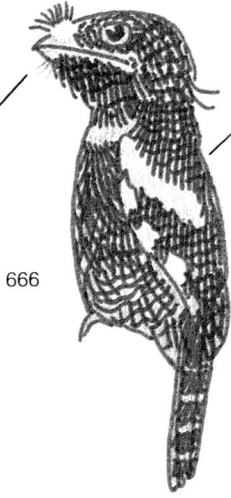

666

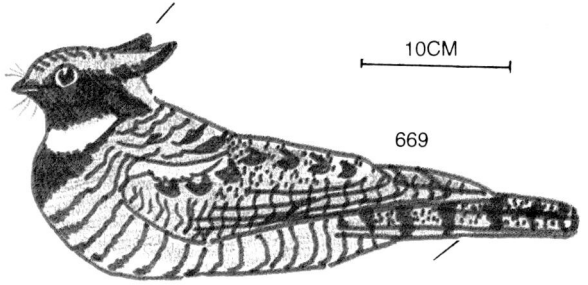

10CM

669

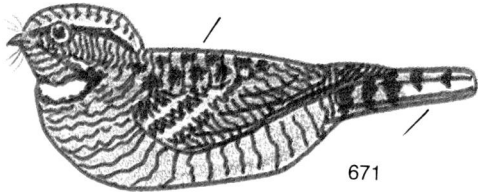

671

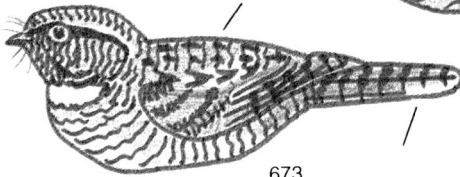

673

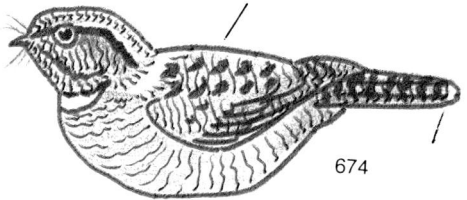

674

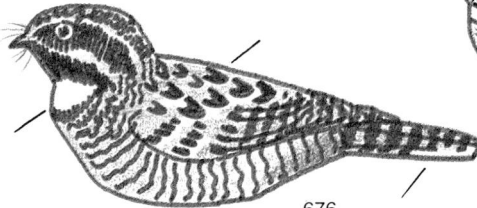

676

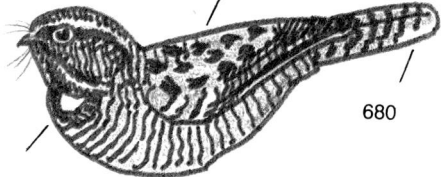

680

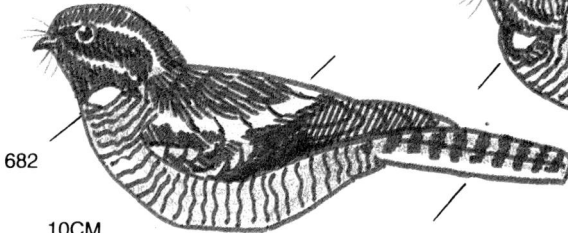

682

10CM

10CM

Plate 36

685

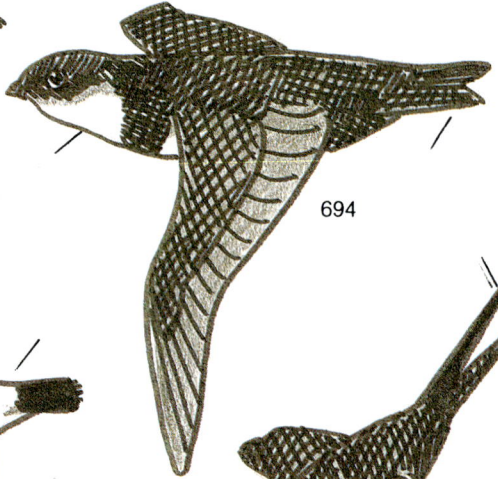

694

692

709

703

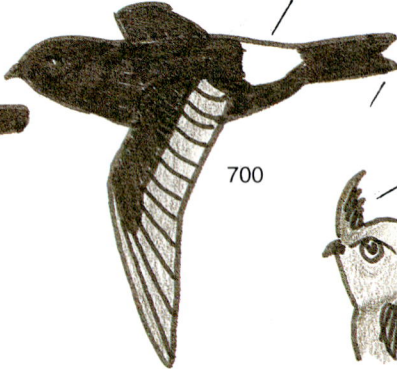

700

709

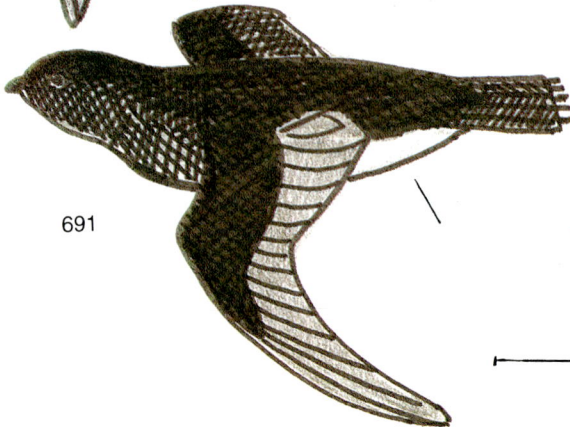

691

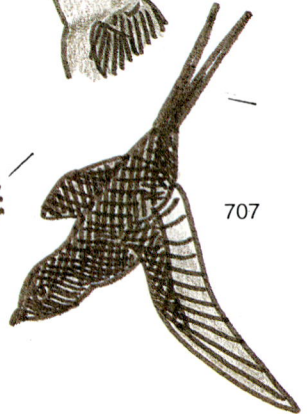

707

10CM

Plate 37

719

725

727

723

730

740

710 ♂

739

736

710 ♀

10CM

Plate 38

744

750

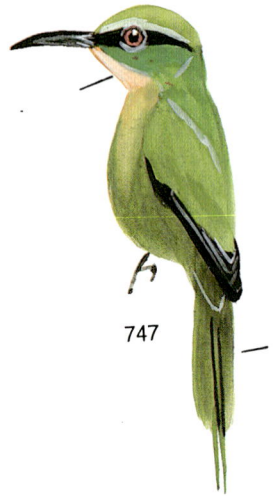

747

10CM

748

753

764

754

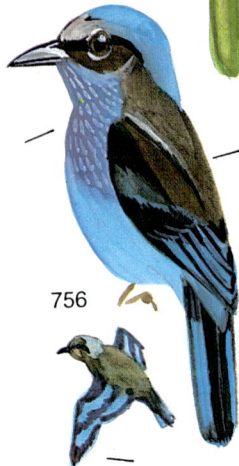

756

759

Plate 39

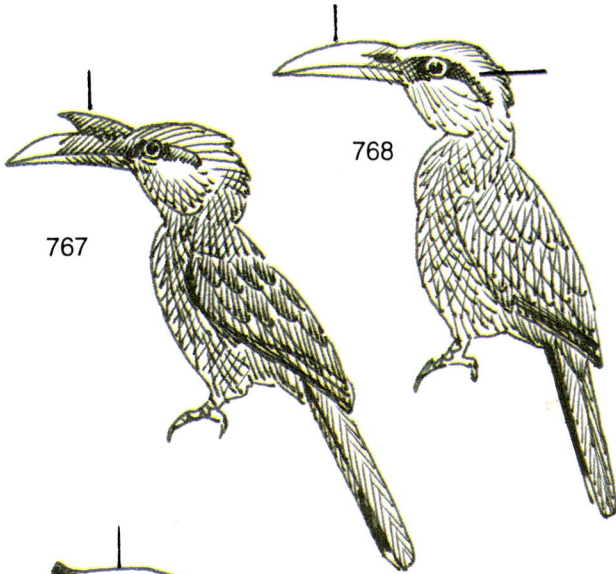

767

768

775

776

50CM

Plate 40

781

785

790

792

799

796

804

852

♂
856

856 ♀

830

816

♂

808

♀

♂
847

♀

10CM

Plate 41

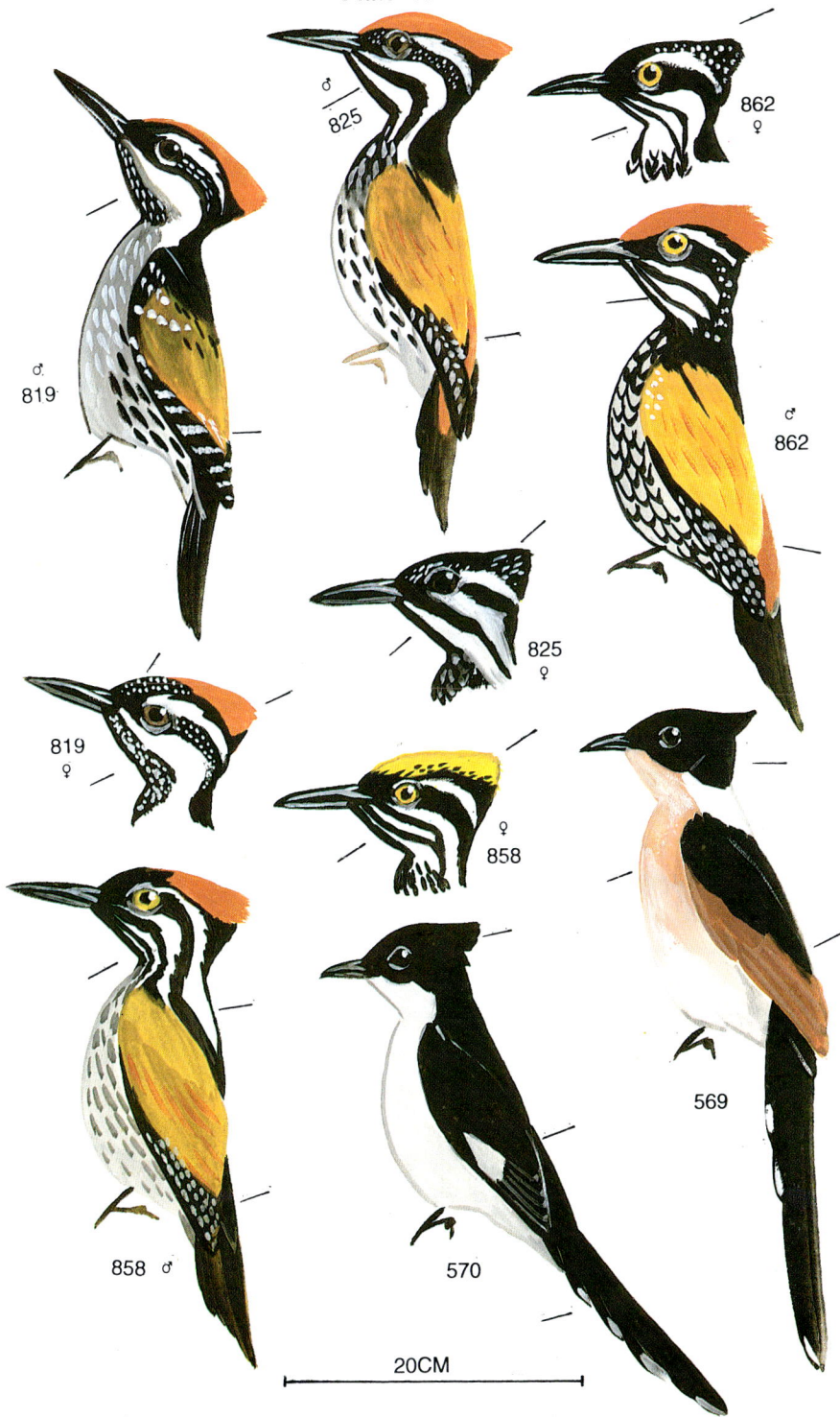

♂
825

862
♀

♂
819

♂
862

819
♀

825
♀

858
♀

569

858 ♂

570

20CM

Plate 42

872

874

877

878 ♀
878 ♂

882

886

907

901

1868

1852

1854

1863

1870

1861

1857

10CM

Plate 43

910

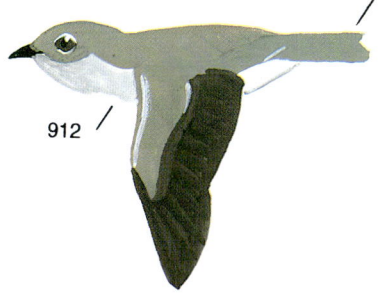

912

913

922

914

930

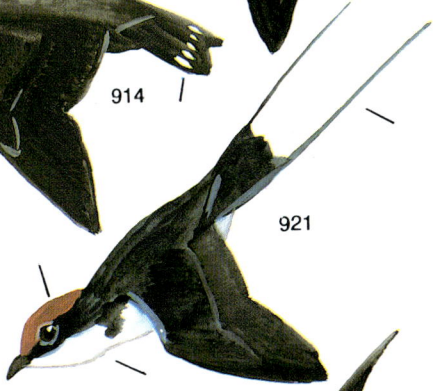

921

925

916

919

10CM

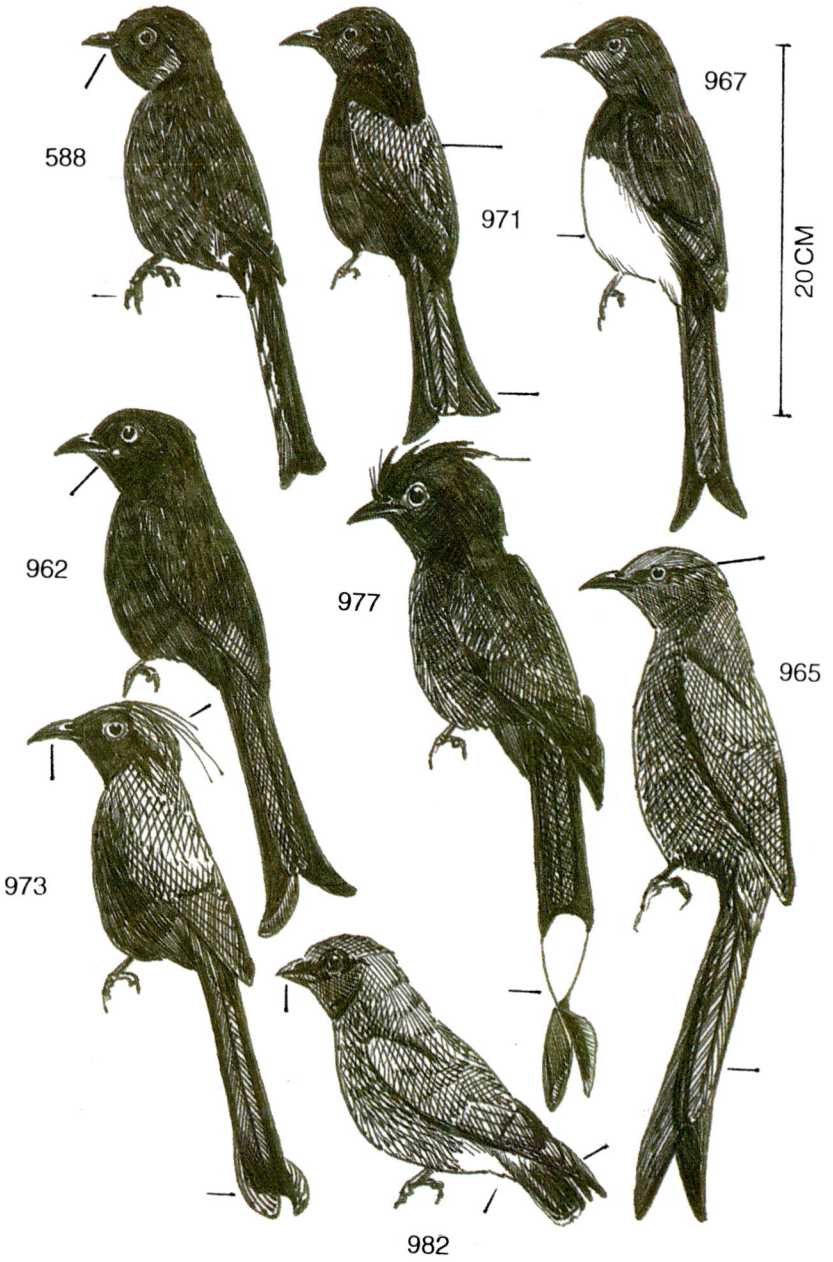

Plate 44

588

971

967

20 CM

962

977

965

973

982

Plate 45

987

988

996

996

994

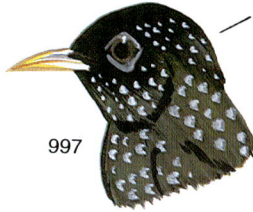

997

1002

1110

1006

1008

1016

12CM

Plate 46

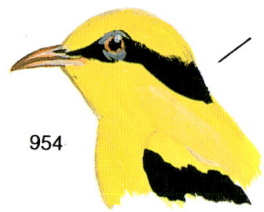

♀
953

♂
953

954

959

♂
1109

1109
♀

1104

♂
1107

♂
1100

1100
♀

1107
♀

10CM

Plate 47

1057

1049

40CM

1036

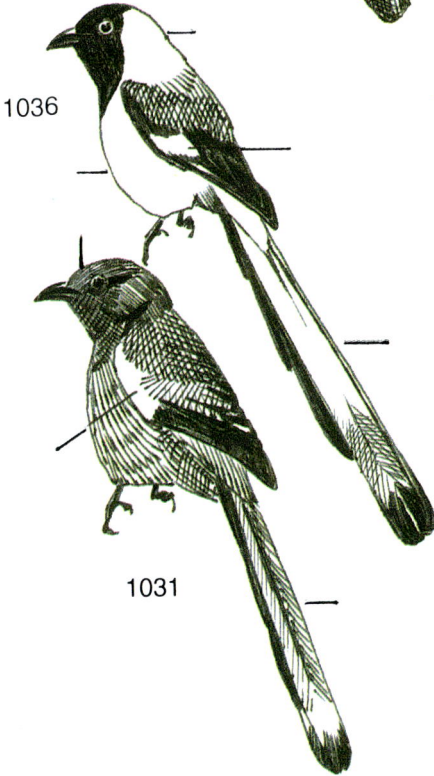

♀

590

1031

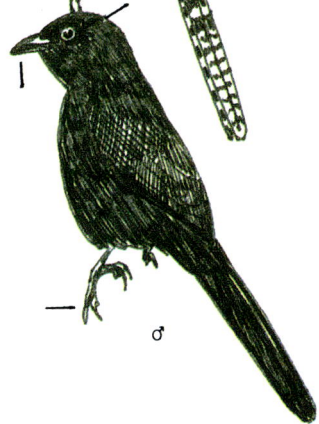

♂

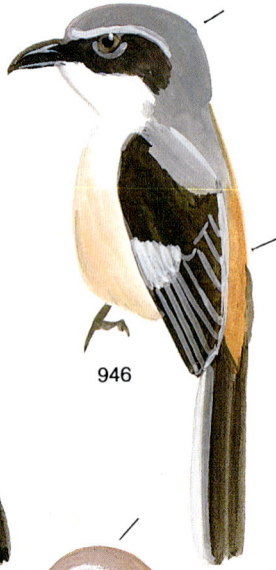

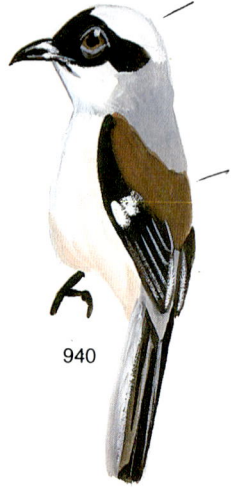

Plate 48

933

946

940

949

943

1063

1068

1070

10CM

Plate 49

♂
1072

1072
♀

1079
♂

1079
♀

1065

1077

1089A

1096
♂

1096
♀

♂
1081

♂
1085

♂
1089

1081
♀

1093
♀

♂
1093

10CM

Plate 50

1114

1116

1138

1143

1135

1123

1149

1127

1120

10CM

PLate 51

1173

1309

1307

1287

1254

1267

1262

1259

1258

10CM

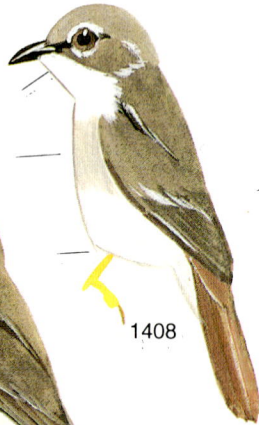

Plate 52

1154

1219

1231

1224

1389

1408

1407

1411

1409

8CM

Plate 53

♂ 1421
♀ 1421
1427
♂ 1440
♀ 1440
1445
♂ 1435
♀ 1435
♂ 1446
♂ 1442
♀ 1442
♀ 1446

10CM

Plate 54

1448

1465
♀

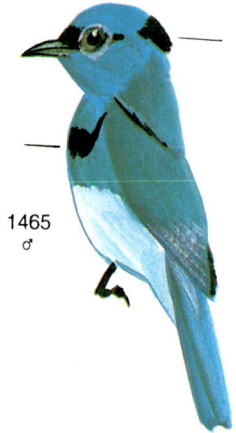

1465
♂

♂
1460

♀
1460

♂
1460

1458

1451

10CM

Plate55

♂
1496

1498

1543

1545

1503

1506

1511

1521

1535

1546

1517

1549

1550

5CM

Plate 56

1556

1557

1567

1565 ♂

1565 ♀

1575

1578

1579

1590

1581

1562

1602

1606

1601

1933

5CM

Plate 57

WING FORMULAE OF WARBLERS

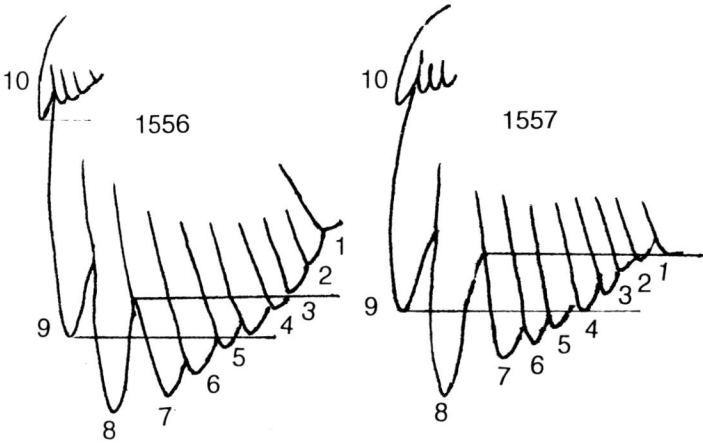

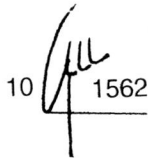

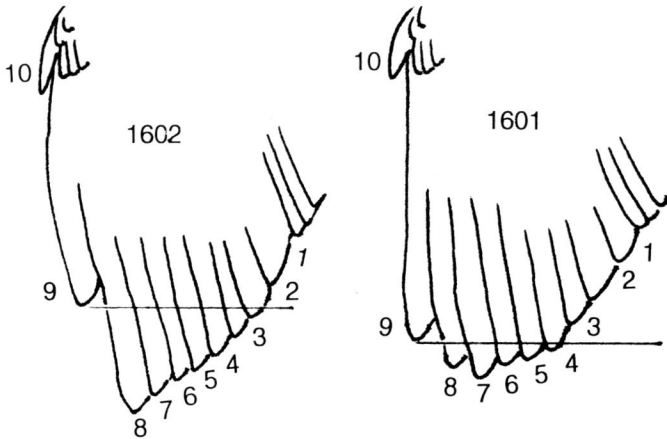

Plate 58

1637

1643 ♂

1643 ♀

1647 ♂

1647 ♀

1645 ♂

1645 ♀

1650 ♂

♀ 1650

1697 ♂

1672 ♂

1672 ♀

1700 ♂

1697 ♀

1710 ♂

1710 ♀

1720 ♂

1720 ♀

6CM

Plate 59

♂ 1723

1723 ♀

1726 ♀

♂ 1726

1661

1665

1728

1742

♀ 1731

♀ 1732

♂ 1731

♂ 1732

1734

1733

♂ 1753

♂ 1748

♀ 1748

15CM

Plate 60

1794

1798

1810

1838

1830

1874

1878

1875

1885

1884

1886

1881

1891

867

10CM

Plate 61

1899

1902

1892

1909
♂

♂
1917

♂
1917

1917
♀

1909
♀

♂
1929

1929
♀

♂
1907

♀
1907

1931

1911

♀
1911

♂

1911

♂

1911

10CM

Plate 62

♀ 1938

♂

1949 ♂

♀ 1957

♂ 1957

♂ 1962

♀ 1964

1964 ♂

1961 ♂

♀ 1961

♀ 1962

1965

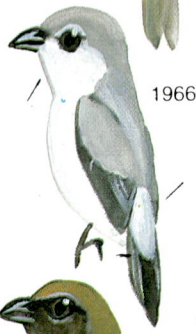

1966

1968

1971

1971

1974

1978

1974

1978

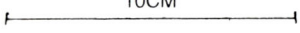

10CM

Plate 63

♂
2010

2010
♀

♂
2043

♂
2044

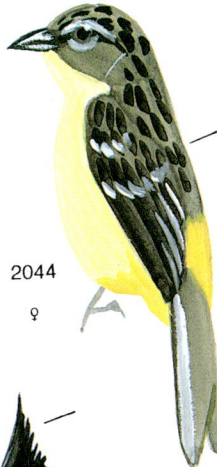

2044
♀

2043
♀

2050

♂
2060

2060 ♀

10CM

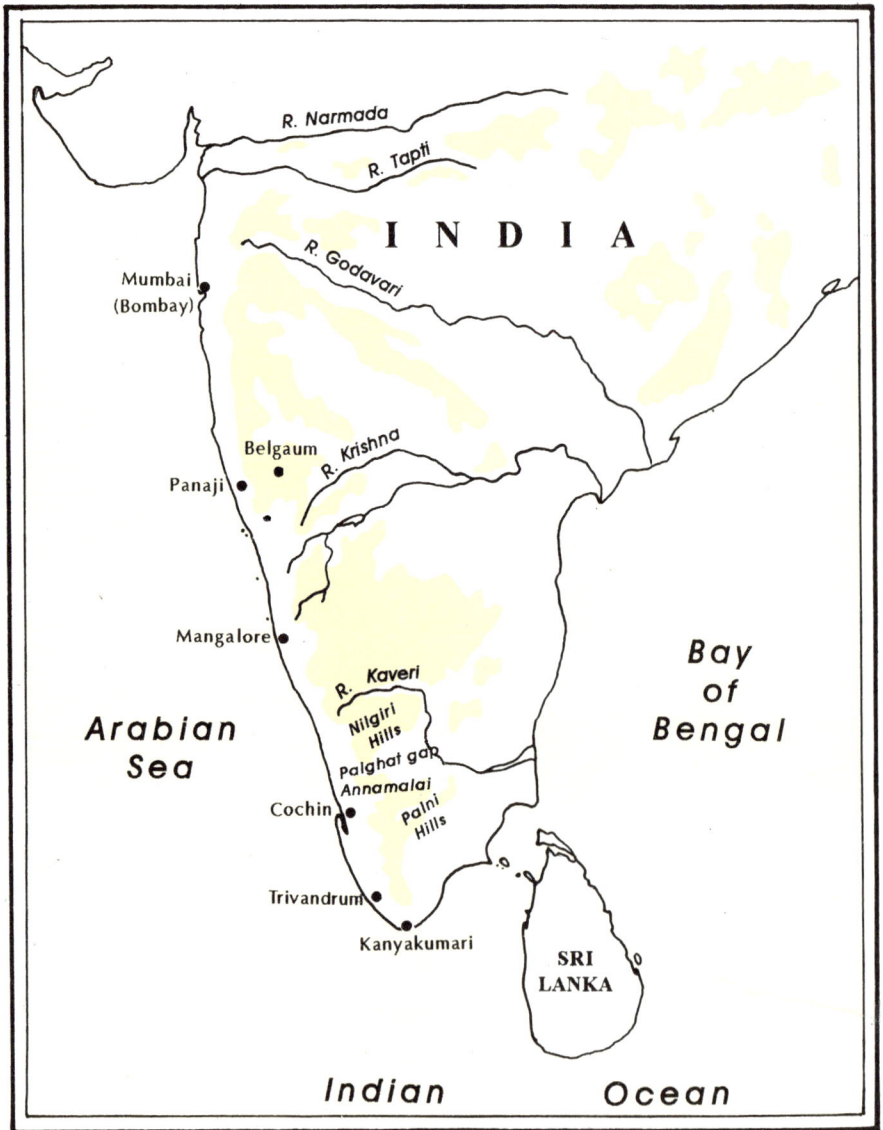